SpringerBriefs in Philosophy

More information about this series at http://www.springer.com/series/10082

Fabio Grigenti

Existence and Machine

The German Philosophy in the Age of Machines (1870-1960)

Springer

Fabio Grigenti
Università degli Studi di Padova
Dipartimento di Filosofia Università degli
Studi di Padova
Padova, Italy

ISSN 2211-4548 ISSN 2211-4556 (electronic)
SpringerBriefs in Philosophy
ISBN 978-3-319-45365-1 ISBN 978-3-319-45366-8 (eBook)
DOI 10.1007/978-3-319-45366-8

Library of Congress Control Number: 2016949415

This Springer imprint is published by Springer Nature
The registered company is Springer International Publishing AG Switzerland

Contents

Introduction

The aim of this work is to provide a partial, preliminary account of the findings of a much more far-reaching investigation into the relationship between technology and philosophy conducted by the author in the context of two international research projects funded by the European Union.[1]

As part of this research effort, the author aimed to apply the methods typical of the historical-philosophical approach to a setting – relating to the issues posed by the enormous development of the *techno-sciences* – that has come to be virtually monopolized by fields of study such as moral philosophy and applied ethics. Contrary to expectations, it soon became clear that – even in research areas apparently far removed from the classic topics of the history of thinking – there are advantages to be gained from applying the historian's skills to revive a tradition in the world of philosophy that, right from its early days, has often taken an interest in the problems of all things technical.[2]

Contemporary thinking is clearly no exception, and so much attention has been paid to this topic in recent times that it has sometimes seemed to exclude all others. The present study aims to step back a little from these recent developments, as concerns its starting point at least. Today's "philosophy of technology" and the related

---

[1] EPOCH (*Ethics in public policy-making: the case of human enhancement*), an international partnership forming part of the European Union's 7th Framework Programme for Research and Technological Development (in which the author served as scientific coordinator of a research project on human enhancement), and SYNTH-ETHICS, which focused on the philosophical, ethical, and regulatory implications of synthetic biology (the author being concerned with the related epistemological aspects).

[2] See, for instance, the work by Giuseppe Cambiano (Cambiano 1971), who explains how Plato's continuous references to techne in his works were decisive in the very constitution of philosophy. On the Italian scene, but no longer on the specifically historical front, the works by Emanuele Severino (Severino 1979) and Ugo Galimberti (Galimberti 2000) are worth mentioning. For a general historical overview and a selection of texts, see Antonella d'Atri (D'Atri 2008). A special mention also goes here to Prof. Paolo Rossi, whose *I filosofi e le macchine* (Rossi 1962) paved the way to a line of investigation that also inspired the present author. Finally, two other works are indispensable for their utility and broad scope, i.e., Anton Hermann and Carl Schonbeck (Hermann and Schonbeck 1993, et seq.) and Ian Mcneil (McNeil 1989).

historiographical reconstructions are dominated by the idea of technology as a sort of compact, totalizing phenomenon with no internal distinctions, something that can be interpreted as a sort of general "world view" capable of pervasive effects and invariably classified as a "hazard". This is a historiographical hypothesis on which Heidegger's great lesson has had a huge influence, but – without detracting in any way from the "greatness" of Heidegger's contribution – the attentive historian cannot fail to see that his is just one testimony among others.

If we take a closer look at technology with an open mind (without first reading what the philosophers have already said on the matter), it becomes evident that we are looking at an area of such complexity that it can hardly be encompassed by such a generic name. "Technology" does not exist, neither as an essence, nor as a set of entities with a common denominator. If anything, there may be many different "technologies". In our day-to-day experience of the technical world, we all have to do primarily with "objects", "tools", "machines", and "devices" that can be used according to different rules and within clearly defined contexts of relations, interests, and expectations. Driving a car, and thereby reducing the distance from one place to another with the aid of a combustion engine, is a very different matter from sending an email while sitting comfortably in front of a computer. These two actions differ particularly in the way in which they can induce changes in our attitude to and relations with our living environment.

Of course, there may be points of view from which the combustion engine and the computer can be seen as one and the same thing, but taking such an approach undeniably fails to accurately grasp the general meaning of specific phenomena that always occur as a result of technological innovations. The extent of the changes induced by the latest media technologies in our way of experiencing the world can hardly be expressed effectively by looking at the effects of the invention of the motor car engine. The philosophers' analysis of the artificial worlds will have to cope increasingly with differential considerations and expect results that are not immediately generalizable. This is essentially what the future has in store. As for the past, the idea of taking a plural view of the fields of technology prompted the choice of a specific topic – that of the *machine* – as a starting point for approaching the pages written by contemporary German thinkers particularly interested in the topic of technology.

The reader might object that the meaning of the term "machine" has just the same "indeterminate" quality as we attributed earlier to the term "technology", but "machine" is used here (together with such synonyms as mechanism, apparatus, machinery, and so on) to mean a fairly clearly defined "object" that appeared on man's living horizon a little over two centuries ago, and it has continued to have profound consequences on our way of inhabiting the Earth ever since.

The term "machine" is used precisely to mean that complex material device assembled in the last quarter of the eighteenth century as a result of the definitive modern refinement of certain fundamental technologies, i.e., metallurgy, precision mechanics, and hydraulics. The "machine" discussed here arrived on the scene of man's history when the processes of spinning and weaving were entrusted to semi-automatic means; when the water wheels used in mills, hitherto always made of wood, were supplanted by the metal levers of the steam engine; and especially when the steam engine was connected to the weaving frames, to the metalworking hammers, and to other machines used to manufacture other machines … in an endless reiteration of assemblies and applications, the enormous outcome of which is what subsequently came to be described as "mass production".

The philosophers discussed here were also dealing with the type of machine described above. More importantly, when they speak of technology as a general phenomenon, they are actually thinking of precisely this specific object, which goes to show that any analysis aiming to define "essences" is ultimately always obliged to refer to "accidents" that can only be confused with universal matters because of their massive, but always nonetheless contingent, invasiveness.

Chapter 1
Epistemological Premise: The Promethean Protocol

Abstract In this chapter are discussed: (a) The basilar concept of promethean protocol as succession of steps – obtaining the raw material, softening, modelling, firing, refining and firing again – clearly form a sequence in which none of the single stages can change places without partly or wholly jeopardizing the quality of the end product; (b) a brief history of the different "gestural theories"; (c) the process of machine's expansion during the industrial revolution. The new place of machine: after thousands of years in which the machine had occupied a modest place in man's world, it suddenly came to the fore, becoming the leading character in the dominant practice of our times: the mechanistic mass production.

Keywords Promethean protocol • Work • Hand • Gesture • Network • Skill • Machine industrial revolution

The term "technology" is used to indicate sets of objects that sometimes differ considerably from one another. Generally speaking, the fields of the various technologies converge into a broad domain with uncertain boundaries that we might call artificial. "Technical" has always been seen (and understood) as opposed to what is "natural". In actual fact, the majority of the things, environments and devices that make up the world of human beings are the product of a more or less complex processing activity conducted in a manner that bears little or no resemblance to the processes by means of which Nature produces her creations.

Man turns to the physical world to obtain materials and examples of forms on the basis of which he constructs his artefacts, but even such a simple object as a cup is the product of a series of operations that can be completed entirely within the already broad horizon of human technical expertise. First the raw material must be found, which is only available in certain places and it can only be extracted with the aid of specific implements. Then moisture must be added to the clay to soften it, and it must be allowed to settle in suitable containers until it turns into a ductile, malleable paste. It is only then that it can be fashioned by a human hand. This is where the potter's wheel becomes very important, as the means that enables the manual movements of the ceramic worker to adjust the size and shape of the object. Then comes a firing step, at a given temperature and for a given time, depending on the particular type of process underway. Exposed to heat, the ceramic object begins to take on its final features: the water content evaporates, and the firing process con-

F. Grigenti, *Existence and Machine*, SpringerBriefs in Philosophy,
DOI 10.1007/978-3-319-45366-8_1

verts the clay into a strong, hard substance. Once it has cooled down, our simple cup can be decorated and enameled, then fired again to make the paints applied to its surface blend into the underlying material, resulting in a shiny, watertight coating that gives the object its final appearance.

This succession of steps – obtaining the raw material, softening, modelling, firing, refining and firing again – clearly form a sequence in which none of the single stages can change places without partly or wholly jeopardizing the quality of the end product. It would be impossible, for instance, to reverse the order between the first and second firing because both the consistency of the unfinished cup and the characteristics of its decoration would be impaired: the cup would be more fragile, less waterproof, and certainly more ugly.

In every field of technology, the chain of operations that follow one after the other until the product has been completed acquires a normative value or, if you will, it serves as a *protocol.* Protocols do not establish generic guidelines that may be variously interpreted; they are strict sequences, and the stages involved must be repeated in the order that experience has shown generate the best results. In addition to the quality of the object being manufactured, this also enables successful techniques and their relative invariability to be passed on, to become part of a cultural heritage. Considered in its essential aspects, ceramic technology has remained much the same ever since ancient times. Even today, it continues to follow the same strict protocol with only minimal differences, especially when compared with the changes that other technologies have undergone.

The same certainly cannot be said of another technological protocol, steel-working. A close relative of ceramics (probably born quite soon after the art of firing clay), it shares with the latter the need to use fire to melt the metal, and especially the problem of how to manage the calorific power released by the furnace. From this point of view, ceramics and steel-working can be seen as applications of one and the same general operating framework that, since it rotates around the use of fire, we might call the *Promethean protocol.*

We are constantly reminded of the importance that man has always attributed to his having learnt how to manage fire: in myths and stories it seems to have always been taken for granted that a very close, specific link exists between man's successful control of this resource and his first great leap forward towards civilization. It is thanks to fire that we have not only improved our physical living conditions, heating our homes and cooking our food, but we have also acquired such immaterial goods as writing and religious beliefs, which have a direct link with the power of heat that might not be immediately obvious. We may be less aware of how our ability to light, feed and exploit the power of a fire can have contributed to the development of the highest expressions of human spirituality because we are used to perceiving the latter more as the causes than as the effects of technological innovation. This attitude stems from an incomplete consideration of what the initiation of a whole technological protocol really requires. In the case of ceramics and steel-working – in what we have called the Promethean field of technology – alongside the series of technical operations established by our protocol, we must also consider two other aspects that are sometimes overlooked: the use of certain manual gestures characteristic of human hand-eye technicism; and the activation of a specific creative function that generates and preventively shapes the manufactured object in the mind's eye.

For several decades, the topic of "technicity" and the type of activity that the term expresses have been attracting a wholly new type of attention. "Handiwork" and its product (the "work") have been acknowledged as constitutive elements of the human condition as part of a general rediscovery of the value of manual action.[1] In the recent debate we can distinguish several interpretative models that will be adopted as representative of the state of the art for at least three basic reasons: (a) they are historically significant because they stem from views that have traditionally been reiterated over time; (b) their goal is to clarify the sense of human technicism in general; and (c) they are models of interdisciplinary provenance and are capable of generating effects of sense in different epistemic settings. These models are outlined below.

- *the architect model, on the dichotomy between "intelligence" and mere "manual execution"*[2]: the former is attributed the ability to visualize the object in advance, to design it and plan its execution; the latter has the dexterity needed to construct the object, but this is seen as a "blind", merely mechanical activity. According to this model, the whole technological action is guided by the mind, by a rationality that foresees what the hand or instrument need to do, while the latter have no virtue in themselves, they can only operate to serve the idea. This model has ancient roots (by Plato and Aristotle), but became better established in more modern times with the advent of the new mechanistic science of nature.
- then there is *the instrumentalist theory, on the original bond between hand and tool***:** in this case the prime agent is the hand, with its structural features and capacity for movement. The hand projects its own forms and features onto the tool, guiding the latter's movement on the grounds of its own "virtue", refined by practice. Starting from this assumption, it is easy to see how we can come to hypothesize the "obsolescence of manual actions", interpreted as a process of loss and decentralisation of human capabilities, which are gradually incorporated in the new subject of technological action – the *machine*.[3] The movements

[1] See, for example Arendt, Hannah. 1956. Shepresented categories such as the *vita activa* (the title she preferred) in the modern world. She defined the three human activities as labor, work and action, with two mutually exclusive spheres: the political and everything else. For us is particularly important the concept of "work" in opposition to the other similar activity: the "labor".

[2] For a wide recognition about this perspective see Mitcham 1979, who argues that this point of view becomes dominant with the radical transformation in Western cosmology ushered in by such figures as Galileo, Newton and Descartes.

[3] About this perspective see further, chapter II. André Leroi-Gourhan's theory of the co-evolution of manual and intellectual activities presents a radical contrast to Chomsky's contemporary mentalist theory of language that get back to Cartesian rationalism with its oppositions between man and beast, and between body and mind. On the contrary, Leroi-Gourhan offers an integrated approach to human evolution: gesture and speech are regarded as twin products of an embodied mind that engendered our technical and social achievements. He asserts that the liberating of the hand from locomotion led to the liberating of the face from prehension, thus creating the duality of instrument and symbol whereby human beings physically and mentally grasp the world in which

of a hand using a tool or operating a machine would always entail a process of *externalisation* (organic projection or estrangement, depending on a given author's point of view) by virtue of which typically human forces, competences and characteristics are displaced and objectivised outside the body at work. Leroy Gourhan (1964) sees the "omission of the gesture" from the context of sensory and cognitive participation in the manual action, and the incorporation of gesture and tool in the mechanical operation as two facets of the loss of man's natural technicity implicated in modern-day technological development, especially in information technology;

- yet we have the *skill/practice theory* that could be defined as the *theory of the original relationship between body and environment.* Ingold identified the concept of *skill* as the point where the nature/culture, organic/social, mind/body, art/technology antitheses intersect and are overcome. Starting from the idea of the body as a flexible, plastic organism, i.e. not culturally or biogenetically encoded *a priori*, Ingold (1999) claims that – also in its technological actions – the body *enfolds* particular sequences of practices, habits and skills within its anatomy, musculature and neurology that are biological and social, mental and physical at the same time. These skills include the capacity of living organisms to adopt and consolidate circuits of posture and gestures that, by dint of repeated practice, is memorised and established as a bodily conformation. In all their aspects, technological actions are among these *skilled practices*. In this case, the primary subject is the body, which is not guided by a mind like the tool; instead the body uses tools and produces objects, exploiting an acquired dexterity (*virtus*), based on the recall of previously-performed and successfully enfolded actions (*savoir fair*). This approach has its historical roots in Antiquity (Plato and Aristotle) and significant points of tangency in such authors as Heidegger. It has its strengths (it is not dualistic, it demystifies the role of instrumentality, it paves the way to a global consideration of technological action), and its weaknesses: technicity almost becomes a sort of "essential super-function" or "emerging property" in the organic-perceptive field that humans share with numerous animals (where "intelligence" and "rationality" have no place!?), which governs different ways of behaving and performing – dexterity, mechanicality, art, ethics, knowledge of the world – making them indistinguishable from one another;
- the *Actor Network Theory*, presented in the setting of *Science and Technology Studies*[4] (abbreviate con la sigla STS) is a model that also aims to overcome the traditional dichotomies between subject and object, nature and culture, mind and

---

they live. This original explanation of the evolutionary association between the hand and the face provides a biological basis for cognitive as well as communicative aspects of gesture, with culture emerging as an extension of our anthropo-zoological structure (see Gourhan 1960, Gesture and Speech).

[4] The *Actor-Network-Theory* is particularly difficult to explain and summarize. In relation to our topic, this perspective dissolves the action and the gesture, arriving both to coincide with the network. The problem is to define clearly the concept of network, which appears ambiguous and dual: on the one hand the network is only the structure (or the form) on the other hand she seems the mere process. In both cases it is very difficult to see how the network can replace in itself the com-

body, etc., considering the world-environment as a network of relationships between different entities. Within this network, every node is occupied by an *actant*, a term used to mean any individual or collective agent capable of joining or leaving collective associations with other agents. The very essence of the actant lies in this characteristic of being part of a network; the actant itself has no substance *a priori*. This means that, within the actor network, the actants may be both human and non-human operators (e.g. machines or technological devices) that are interconnected and capable of influencing one another. From this perspective, the essential point is that the action (*agency*) – any type of action – is no longer the use of tools or of any "human skills", but the overall infrastructural *effect* deriving from the fact of being in a network. This element of the action's distribution, that in STS settings is called the *principle of generalised symmetry*, has a number of other consequences (the working capacity and ability to implement an action attributed to non-human entities, socio-material hybridisation, the historicity of things, and reality as a "good construction");
- finally the *theory of expertise as a craft*, that we owe mainly to Richard Sennet (2008), reiterates and relaunches the view of the pragmatists (Peirce, Mead, Dewey and James). The return to the idea of "craftsmanship" – that Sennet intends as a sort of expression of the impulse to "do a job well for its own sake" – is intended not only to restore value to the expert action that is intrinsic in the relationship between head and hand, but also to lend it a certain ethical quality, something that is not just being done on the strength of a given expertise, it is also being done according to a normative criterion of goodness and adequacy.

The state of the art thus appears to be characterised by interpretative models that may all claim to being generally applicable, but they are actually highly differentiated as regards the following issues:

What is the role of the human gesture in technological practices? When human gestures are replaced by mechanical actions, is this an opportunity for growth or an alienation? Who is the subject of the operating circuit: the mind, the hand, the machine, or the network? What role do objects and machines have? What epistemic value does technical-manual action have? Is it intelligent action, a skill of the body, or of hybrid entities, or of the network as a whole? Who takes responsibility, and on the grounds of which values, for the effects of the action, a cognizant subject, a living organism, the master craftsman in his laboratory, a machine, a collective group that takes action with a shared purpose?

This scattered and multifaceted scenario, this plurality of answers is due to: (a) the different theoretical premises supporting the conceptual foundations of the various views; (b) their focusing on a particular type of "technological action" and interpreting it as absolute and generalisable; and (c) their paying excessive attention to the statute and role of the *agents* (rationality, hand, body, actant), and not enough to that of the objects and to that instrumental-mechanical context.

---

plexity of gesture, that is form, process, skill, body movement…and so on. The most important esponent of these widespread theory are Latour 1987, 2005; Callon 1986, Law 1987.

The anthropologists have begun to make it clear[5] that the evolutionary step that generated the *Hominidae* involved several changes in the vertebrates' skeleton, including the verticalization of their spinal column (enabling an erect stature), which meant that their front limbs – no longer needed for support and dynamic anchoring purposes – could gradually become highly perfected means for manipulating objects. The mobility of the human hand, and the remarkable sensitivity and precision with which the fingers can touch, select and retain objects of even the tiniest dimensions, have no equal among the primates. Humans are essentially beings who explore and adapt their world with the aid of their upper limbs. The motions of the ceramic worker shaping a vessel, narrowing and widening it with the palm of his hand, contain the essence of a technical skill natural to our species. It is easy to see that this is due not to any spiritual cause or superior intelligence, but to a given mechanical advance. The adoption of an upright posture gave rise to a mechanical structure capable of doing certain things, which include the set of functions performed by a brain that was able to develop enormously thanks also to the displacement of the point where the vertebral column enters the skull prompted by the verticalization of the trunk. The current position of the occipital foramen in humans undeniably facilitated an expansion in the skull's volume, enabling the development of a large brain mass, particularly developed in the areas that govern functions such as vision and language. Human beings fashion and handle objects by means of a refined visual control system capable of supervising the action circuits, distinguishing between materials, recognizing shapes and instruments. The eye follows the hand's movements, controlling every step, but it is also capable of detaching its gaze and scanning the horizon, seeking, anticipating and preventing. This panoramic capacity of human vision, this ability to be distracted momentarily from the action the hand is performing, probably explains why creative spirit is so important in technological circuits. The object is not simply produced, it is first seen in the mind's eye as an "idea" that can be put into practice.

This abstract modelling of an object acquires a fundamental role in operational chains of Promethean type, as the very name of Prometheus reminds us. Even before he has begun to shape his lump of clay, the potter has already seen the shape of his cup in his mind. This mental picture of his cup appears as an "end" in itself, it has a "purpose" over and above the material result of his actions. The ideal model seems to *embody* the whole process, guiding every single step until the real cup has been completed. It is highly likely that this apparent supremacy of the ability to imagine the result in advance made it possible for the manipulatory function to be considered separately from the ideational design function, giving rise to a separation within the protocol, that was promptly translated into a hierarchy between functions, practices and institutions. The claimed superiority of the mind over the hand,

---

[5] See, above all, A. Leroi-Gourhan (Leroi-Gourhan 1964) who seeks to demonstrate in *Gesture and Speech* that man is simply the product of a development in the phyletic lineage of the vertebrates that adjusted his anatomo-postural situation and enabled him to walk on two feet. The pages that Oswald Spengler dedicates to the matter of the human hand becoming free to handle tools in *Der Mensch und die Technik* (Spengler 1931) had a great influence on the authors discussed here.

and the consequent stable establishment of a "political" preeminence of the former over the latter, constitute one of the first characteristic "abstractions" of Western civilization. This led to the birth of a particular "eidetic" technology – philosophy – the primary function of which became to process immaterial objects (ideas) that, from then on, were intended as the very essence not only of what man produced, but also of everything that he might encounter in the world around him. This passage coincided with the hand being reduced to mere executor of an "ideal" command that took shape not in gestures requiring technical expertise, but in writing and the word.

We shall return to this crucial passage elsewhere.[6] What is worth emphasizing here is that, in order to create a hierarchy, it drew an artificial distinction between what, from the technological standpoint, should be accepted as wholly integrated and equally important. The hand without the eye would be blind, the eye without the hand utterly unproductive. The essential link between these elements of human technicism means that any variation in the usage of the one part triggers changes in the other, thus enabling the mechanism as a whole to do new things. This helps us to understand how the adoption of novel technological solutions has sometimes triggered great "spiritual" revolutions, and how systems of ideas and beliefs have often had the effect of facilitating, or interfering with the consolidation of innovation in the technical field.

The mutual influence of hand and eye has certainly also been at work in the changes that the Promethean protocol has undergone, although both the versions of it considered here (in the fields of ceramics and metalworking) remained substantially unchanged up until very recent times.

For thousands of years human beings have been fashioning ever more magnificent objects thanks to a perfect cooperation between their eyes and their hand movements, and they continue to do so today. What they have not succeeded in improving over time is their management of the energy consumed in the process. The power that a human body can generate is somewhat limited and it can vary considerably from one individual to another. Once exhausted, moreover, it can only be restored after a rather lengthy period of rest. The situation was first improved by the introduction of animal power and by exploiting the energy obtainable from water and wind. Using these forces enabled mankind to develop a basic technological civilization. But their capacity for work remained minimal, and the problems relating to their susceptibility to fatigue were compounded by the difficulty of finding resources, and their inconstant availability. There was a further problem, relating to the time it took humans to improve their metalworking technique; and even when they did learn to fashion metals into extremely useful implements, the latter remained much the same for thousands of years. But a number of decisive changes eventually took place on these elements, the results of which ultimately led to what historians call the Industrial Revolution.

> The Industrial Revolution began in England in the eighteenth century, spread therefrom in unequal fashion to the countries of Continental Europe and a few areas overseas, and trans-

---

[6] Cf. an essay by the present author entitled *La scrittura del corpo* (in press, Padova University Press 2016).

> formed in the span of scarce two lifetimes the life of Western man, the nature of his society, and his relationship to other peoples of the world [...] The heart of the Industrial Revolution was an interrelated succession of technological changes. The material advances took place in three areas: (1) *there was a substitution of mechanical devices for human skills*; (2) *inanimate power – in particular, steam – took the place of human and animal strength*; (3) *there was a marked improvement in the getting and working of raw materials, especially in what are now known as the metallurgical and chemical industries.*[7]

It is a widely-held opinion that the Industrial Revolution was, above all, a technological revolution. If we look carefully at the three above-mentioned areas of activity that it influenced, it is also easy to see that each of them was changed as a result of the incorporation of a particular *machine* in the circuit for performing the activity concerned.

Concerning the first point, it could be said that the introduction of John Kay's flying shuttle in 1733[8] triggered an unstoppable process of successive *substitutions* of the human hand by mechanical means in all the crucial steps of the weaving process. These developments took almost exactly a century to complete, and in 1830 the last of England's "manual" weavers gave up striving to resist the use of the mechanical weaving loom. It took much less time to complete the passage from the use of human and animal power, wind and water to the use of the power generated by the *steam engine*: the version developed by Watt in 1775 was soon being applied in all of the fundamental sectors of industrial production. Finally, every advance in man's capacity to extract raw materials (and especially coal and iron), and to process them more efficiently, was made possible by an endless series of innovations in the means for managing heat (furnaces) and rolling metals.

Within the very short space of a century, all the technological sectors founded on the technical applications of hand-eye coordination and on man's capacity to control natural energy were changed beyond recognition by the introduction of machines at every node in the production process. The basic Promethean protocol was not superseded, the elements involved in the manufacturing process and its general purposes remained the same. Instead, the industries underwent a profound reorganization that, for the sake of simplicity, we might describe as an *expansion of the machine.*

The revision and reassembly of the protocol was immediately experienced as an unprecedented, profoundly revolutionary event. For the first time, an artefact – the machine – was replacing human beings in elementary aspects of the production process, substituting the previously-used tools, which were now removed from the human hand and attached to a mechanical body.

After thousands of years in which the machine had occupied a modest place in man's world, proving useful in humble, unobtrusive ways, it suddenly came to the

[7] DAVID S. LANDES, *The Unbound Prometheus*, Cambridge University Press, Cambridge 2003 (I Edition 1969) p. 1.

[8] On these matters, see the important work by V. MARCHIS, *Storia delle Macchine. Tre millenni di cultura tecnologica*, Laterza 2010; cf. in particular, pp. 139–206. The references at the end of his work also offer a general overview of the studies conducted on the "machine" from a historical and engineering standpoint.

fore, becoming the leading character in the dominant practice of our times: industrial mass production. This moment of glory prompted a much more appropriate definition of the term:

> The name of machine is used for all systems of bodies destined to transmit the work of forces and consequently capable of modifying the intensity of said forces and varying a movement in terms of its intensity, velocity and direction. The variations in the path covered by the moving parts are what make machines particularly suitable for an infinite number of industrial uses.[9]

Machine body, work, transmission of forces, intensity and *speed variations* became the load-bearing lemmata of a new "philosophy of the machine" that spread with the speed of a new language. Though still not accurately encoded, this is the language that describes the real experience of an epoch of which – even now – it is hard to gauge the precise contours and outcomes.

In the following pages, we shall focus on just one, spatially and temporally specific, segment of this philosophy on the topic of the "machine". This concerns certain significant expressions of German philosophy published between the end of the nineteenth century and the middle of the twentieth. Between 1853 and 1870, Germany completed its own "industrial revolution".[10] In no time at all, the country took a great leap towards modernization that it would be wrong to qualify simply as an emulation of the English model. In some sectors, and especially in the metallurgical industry, Germany did not restrict itself to retrofitting the machines arriving from the other side of the Channel. It engaged instead in a remarkable effort of innovation that, by the 1890s, enabled Germans to compete on an equal footing with their British cousins.

This progress was not without its costs. In the course of a single generation, the Germans profoundly and irreversibly changed their country. They did away with a whole series of "constraints" – territorial fragmentation, restrictions on free enterprise, limited financial resources, professional associations – that were objectively posing almost insurmountable obstacles to the development of capital. Such an upheaval in the traditional setup of German society prompted lively debate.[11] Politicians, economists, sociologists, philosophers, and men and women of letters joined the discussion on the value (or non-value) of the technological modernization that the country underwent in the years between the end of the nineteenth and the start of the twentieth centuries.

[9] M. C. LABOULAYE (ed.), *Dictionnaire des arts et manufactures*, Librairie de L. Comon, Paris 1853, p. LXV. On general and specific aspects relating to the "century of machines", see above all: L. MUNFORD, *The Myth of the Machine, Technics and Human Development*, New York 1967 (trad. it. *Il mito della macchina*, Il Saggiatore, Milano 1969); A. DE PALMA, *Le macchine e l'industria da Smith a Marx*, Einaudi, Torino 1971; G. B. DYSON, *Darwin Among the machines: The Evolution of Global Intelligence,* Basic Books, New York 1997.

[10] See again DAVID S. LANDES, *The Unbound Prometheus*, p. 256.

[11] On this matter, see above all: T. MALDONADO (ed.), *Tecnica e cultura. Il dibattito tedesco tra Bismarck e Weimar*, Feltrinelli, Milano 1979.

It would be impossible to give an account here of all the different opinions, or of all the topics that came up, but two general characteristics of the debate cannot fail to stand out. One is the extremely penetrating and competent way in which the changes induced by the expansion of technology were described. The other concerns the awareness that a painful transition had taken place and, although this transition was implemented in the name of progress, it was immediately interpreted in terms of a "loss", a "hazard", and the "decline" of an order that had hitherto always rotated around human and cultural values.

In a sense, these same elements were also identifiable in the analyses of the philosophers, who were among the first to take the floor in the debate and were certainly among the last to hand over to other speakers. As we shall see later on, the philosophers continued in their critical appraisal of technology up until long after the Second World War. Adopting a method rather like an opinion poll, that starts with a specific question instead of trying to describe a whole picture, we shall take a look at the reflections of several, very different writers who analyzed the topic of the machine at various times. Then, on the strength of their comments, we shall try to sketch a general picture.

Chapter 2
Karl Marx – From Hand Tool to Machine Tool

Abstract In this chapter is presented the analysis that Marx leads about the transition from the mere tool (Handwerk) to the machninery (Maschinerie). According to Marx the consequences of this transfer of the manual implement from man to machine are: (a) the machine becomes the active, super powerful, but completely depersonalized subject in the production process; (b) the human body has to learn new actions that are dominated by automatic repetition; (c) the introduction of particular type of relationship between humans and machines: in this new "technological" master - slave relationship it is no longer a case of the instrument serving the man, but of the man serving the machine; (d) the new aspect of machinery: the enduring symbiont.

Keywords Marx • Toll • Machinery • Subject • Object • Depersonalization • Repetition • Master-slave relationship

According to Marx, the introduction of machines in industry was due not to the fact that using mechanical means to do some of the work could reduce the burden of fatigue for human beings, but to the discovery that a machine can be a formidable means for producing a surplus of goods. It shortens the part of the working day that the worker uses for himself, consequently prolonging the part of the working day that he makes available free of charge to the capitalist.

But this merely economic usage of the machine is not justifiable per se. Marx sees it as being founded on the particular nature of the "machine as a means", on what we might call its "inner structure". Unlike the passage that occurred with the advent of manufacturing, which was made possible by using the human workforce in a new way, the revolution that occurred in the industrial world with the birth of mass production was based exclusively on the new form acquired by the *means* that, from mere *tool* (*Handwerk*), had become *machinery* (*Maschinerie*).

Marx's analysis of the historical significance of this transition is ambivalent. On the one hand, he says it is impossible to draw "hard and fast lines of demarcation"[1];

[1] K. MARX, *Das Kapital Bd I*, in *Karl Marx-Friedrich Engels Werke*, Band 23, Vierter Abschnitt, Dietz Verlag, Berlin 1968, p. 391 (Eng. trans., by B. Fowkes. *Capital. A Critique of Political Economy*, I, ed. By E. Mandel. Penguin Books, London p. 392). There has been a renewed interest in the figure of Marx of late, also in relation to the topics discussed here. Among others, it is worth

F. Grigenti, *Existence and Machine*, SpringerBriefs in Philosophy,
DOI 10.1007/978-3-319-45366-8_2

on the other, he pauses (not without some degree of admiration) to describe the revolutionary details of that "productive organism that is purely objective"[2] represented by the system of machines that the worker finds before him - not as a tool for him to use, but as a complex automaton that he must serve. This ambivalence in Marx's analysis cannot be seen as a shortcoming: it stems from the author's necessarily oscillating gaze when he examines the particular "historical context" of technological innovation. The new machine is the outcome of old machines being reassembled and incorporated as component parts of a new structure. Sometimes these single components are enlarged or miniaturized, but their shapes and functions are not new, they had already been configured in previous devices. In a given technological field, the latest machine to be built can never really be described as "revolutionary" if, by this term, we mean something that is "in discontinuity" or "incommensurable" vis-à-vis something that existed in the past. This does not alter the fact that such a machine can prompt enormous changes, the extent of which might even be amply underestimated were they to be described as a "revolution". In the description that he provides in *Capital*, Marx conveys his awareness of this duplicity of machinery. In what sense, he wonders, are the new machines used in mass production distinguishable from the craftsmen's traditional implements? Looking at the inner structure of each mechanical assembly suggests a general layout consisting of three parts: drive, transmission and machine tool, or working machine. The drive element induces the movement of the assembly as a whole. It can be powered by a human body or an animal, or by natural sources of "energy" such as water and wind. In its extreme stage of development, this propulsive power is supplied by other machines that convert the power of steam or electromagnetism to guarantee a virtually constant supply of the energy previously provided by external forces. The transmission element comprises a set of components such as wheels, belts, pulleys and shafts, that transfer and distribute the power to move the machine tool, make it change direction, or vary its speed. The third element is the working machine proper, the *raison d'être* for the whole mechanical assembly: it holds and processes the workpiece thus completing the production process. According to Marx, the huge changes that occurred in industrial production methods towards the end of the eighteenth century were triggered by improvements made to this third and last part of a machine. He described the structure of the working machine very precisely:

> On a closer examination of the working machine proper, we find in it, as a general rule, though often, no doubt, under very altered forms, the apparatus and tools used by the handicraftsman or manufacturing workman; with this difference, that instead of being human implements, they are the implements of a mechanism, or mechanical implements. Either the entire machine is only a more or less altered mechanical edition of the old handicraft tool, as, for instance, the power-loom, or the working parts fitted in the frame of the machine are

---

mentioning E. MICHAEL, *Kapital und Technik*, J.H. Röll, Dettelbach 2000; J. VIOULAC, *L' époque de la technique; Marx, Heidegger et l'accomplissement de la métaphysique*, Presses Univ. de France, Paris 2009; A. BRADLEY, *Originary Technicity. The Theory of Technology from Marx to Derrida,* Palgrave Macmillan, Basingstoke 2011.

[2] Ivi, p. 405 (p. 407).

> old acquaintances, as spindles are in a mule, needles in a stocking-loom, saws in a sawing-machine, and knives in a chopping machine, ... *The machine proper is therefore a mechanism that, after being set in motion, performs with its tools the same operations that were formerly done by the workman with similar tools.*[3]

The working part of the machine therefore does not use different processing tools from those used by human hands. The same tools (albeit with some changes in their dimensions) are attached to the machine. So, if we consider the machine as a whole, including its drive and transmission elements, we could say that there is nothing innovative about it. Its inner structure simply reiterates that typically human technicism represented by the link between a man and his implement. It is only by virtue of a *repetition* - that must be as conservative as possible in order to function - that the machine can replace the craftsmen. The machine dispossesses the worker of his tools, and of the confidence with which he used them, taking his place in the final stage of the production process, the one in which the object is given the required *form*.

The essence of the process of mechanization in industry lies, according to Marx, in this handover, which was immediately followed by the related process of industrialization, implemented by means of a massive multiplication of the number of tools capable of working simultaneously on the holder of the same machine. A human being can only work effectively with one tool at a time. A machine can operate simultaneously with a far greater number of tools. Even in Marx's time, people were amazed when they saw that even the less innovative spinning machines could work with 12 or 18 spindles at once, whereas even the most capable human worker could only cope with one.

The machine brought two elements together in the same assembly: an exact repetition of the *form* of human implements (which corresponded to the shape of the product in reverse); and the chance to go beyond the capacity for work of the organic body of the human worker. The type of production previously assured by a human hand holding a tool remained the same. The machine performed the same spinning and stitching actions. But the machine allowed for the number of tools at work per unit of time to be multiplied, and the rate at which the single operations were performed could be increased. The machine could work like a sort of giant human worker equipped with a huge number of hands, all capable of working at a much faster rate than a human worker, however dexterous he might be.

The essence of the machine lay in the merely "instrumental" conception of the technical means being abandoned for good. As Marx put it:

> The machine, which is the starting-point of the industrial revolution, supersedes the workman, who handles a single tool, by a mechanism operating with a number of similar tools, and set in motion by a single motive power, whatever the form of that power may be. *Here we have the machine, but only as an elementary factor of production by machinery.*[4]

[3] Ivi, p. 394, pp. 494–495.

[4] Ivi, p. 396, p. 497.

The passage from the hand using the tool to the latter being attached to a mechanical holder initially left the worker serving just two purposes: to supply the energy needed to drive the machine's movements and to visually monitor its performance. Marx makes the point that, in actual fact, the ever larger size of the working machines and the increasing number of tools operating simultaneously very soon made it necessary to develop larger drive means, that in turn demanded quantities of energy that neither humans nor animals could deliver. All animate beings were thus definitively expelled from the operating circuit of the machinery, and human beings were ultimately only needed in a "supervisory" role, as mere outside observers of a chain of operations that the machine could now perform entirely unassisted:

> As soon as tools had been converted from being manual implements of man into implements of a mechanical apparatus, of a machine, the motive mechanism also acquired an independent form, entirely emancipated from the restraints of human strength. Thereupon the individual machine, that we have hitherto been considering, sinks into a mere factor in production by machinery. One motive mechanism was now able to drive many machines at once.[5]

According to Marx, it was the transfer of the tool from the human hand to the machine that prompted an increase in the power needed to make the whole apparatus move. It was consequently the quality of the work and the intensity of the repetitions per unit of time that led to the steam engine being connected to a mechanical loom, not the other way around. Watt's discovery alone would have failed to transform the industrial process, had it not been possible to transform the process into a mechanism. The machine is not the worker, but it can replace the worker because it operates *as if* it were the worker. Had this transfer of the manual implement from man to machine proved impossible (as is still the case for some human functions), we would not have witnessed the expansion of the Promethean protocol that lay behind the Industrial Revolution. Already in Marx's writings, this revolution came to appear as the expression of a wonderful process of humanization of the world with the machine installed at its active center. It replaced the worker, but by no means supplanted human technical expertise. Quite the opposite, it focused on copying man's structural features and movements ever more effectively, thereby achieving ever higher levels of performativity.

Like humans, machines can cooperate in the workplace. According to Marx, they can join forces in various ways: as agglomerates of homogeneous machines that are all activated simultaneously, or as chains of different or partial machines that contribute to the end product, each completing a part of the total workload.

In the former case, the whole product is completed by the same machine, by means of different tools all incorporated in the same body. The manufacturing of envelopes for letters once involved the workers completing a precise sequence of operations: one folded the paper with a ruler, another added the glue, a third opened the flap where the watermarking was applied, a fourth added an embossed stamp, and so on. Already the earliest machines for making envelopes could simultaneously complete all these steps, producing more than 3000 envelopes in an hour. A

[5] Ivi, p. 398, p. 499.

production process that was initially completed in a series of successive steps, passing the product from one hand to the next, could now be completed on the same machine equipped with different tools. Multiplying the number of these homogeneous devices being powered by the same engine gave rise to a simple form of cooperation.

The situation is different in other industrial processes. In the woolen industry, for instance (a typical example of a process divided into parts and distributed amongst different workers), the various operations – beating, carding, combing, spinning – are completed by linking together different, dedicated machines, each of which can be seen as an "organ" capable of performing a particular "function". This is a complex form of cooperation that actually represents the translation in mechanistic terms of a system for distributing the workload that already existed in the manufacturing world. The flow chart according to which the product derives from a succession of partial processes distributed amongst different manipulators, each working according to their subjective characteristics (individual capacity, physical form, mutual distance) is reiterated in an objective form by machinery, with the advantage of a greater continuity in the completion of the process as a whole.

Here again, the system of machines imposes an intensification of parameters such as the rate at which each partial process is completed, and the speed with which each machine forwards the material being processed to the next machine. The sequence of actions completed by each machine, and the succession of "handovers" from one machine to the next are still based on the human flowchart previously adopted in the factories, and this human organization provided the basic foundations for the technical cooperation entrusted to the machinery.

This cooperative work required of the machines also relies on the availability of a greater power supply:

> Just as the individual machine retains a dwarfish character, so long as it is worked by the power of man alone, and just as no system of machinery could be properly developed before the steam-engine took the place of the earlier motive powers, animals, wind, and even water; so, too, modern industry was crippled in its complete development, so long as its characteristic instrument of production, the machine, owed its existence to personal strength and personal skill, and depended on the muscular development, the keenness of sight, and the cunning of hand, with which the detail workmen in manufactures, and the manual laborers in handicrafts, wielded their dwarfish implements.[6]

Mass production was made possible by overcoming the boundaries imposed on the work by the technical capacity of the human machine. Human tools based on the link between eye and hand not only have intrinsic limits in terms of their precision, but are also moved by a set of muscles of very limited power that soon tire. They also depend structurally on subjective conditions that make it impossible to predict the mean efficiency of the system.

The machine takes on the mechanical part of the human's job and improves the performance of the single apparatuses, functioning like a more powerful, de-subjectivized version of the human worker. In the way machines cooperate, we can

[6] Ivi, p. 403, p. 504.

see the same law of substitution at work. The republic of the machines does not cancel the form of human work, it repeats its instrumental quality at every single step. But it also speeds up the connections between the successive steps in the production process, depersonalizing the overall movement. What cooperating machines achieve is simply the mechanical integration of collective, multiple and differentiated needs with the different capabilities of each individual component part of the machine assembly, where "individual" is no longer an element that may be dysfunctional for "subjective" reasons; it is always a perfectly efficient component that does exactly the job required of it, for which purpose it has been fashioned.

Given this linking together of different devices, the worker interacts no longer with the single machine, but with the whole factory system. The machinery becomes a sort of enormous automaton consisting of innumerable mechanical parts, all reliant on a single power supply and connected to other non-mechanical parts, i.e. the limbs and eyes of the human operators. Marx describes this new situation in various ways:

> All work at a machine, requires the workman should be taught from childhood, in order that he may learn to adapt his own movements to the uniform and unceasing motion of an automaton. When the machinery, as a whole, forms a system of manifold machines, working simultaneously and in concert, the co-operation based upon it, requires the distribution of various groups of workmen among the different kinds of machines. But the employment of machinery does away with the necessity of crystallizing this distribution after the manner of Manufacture, by the constant annexation of a particular man to a particular function. Since the motion of the whole system does not proceed from the workman, but from the machinery, a change of persons can take place at any time without an interruption of the work.[7]

According to Marx, the consequences of this fully-automated organization of the workload in the factory are as follows.

The machine becomes the active, but completely depersonalized *subject* in the production process.

Humans become not the object of this process, but subordinates of the machine, serving as its *instrument*. While in traditional manufacturing the tool was adapted to the capabilities of the human operator's hand, now it is the body of the worker that has to be adapted so that it can interact efficiently with the machine. The human hand has to learn new gestures that are dominated by automatic repetition.

The standardization of the worker's movements (by virtue of which the steps in the industrial process can be simplified and reduced to a few essential actions in each phase of the process) assures an unlimited interchangeability of the human instrument involved. The worker can be connected equally well to any part of the device, he no longer needs to have any specialization. Taking a more general view, this puts an end to all the hierarchies and all the differences deriving from the personal skills of different workers in the old manufacturing world. Now, as far as the machine is concerned, they are *all the same*.

---

[7] Ivi, p. 443, p. 546.

This introduces and consolidates a particular type of relationship between humans and machines: in this new "technological" master-slave relationship *it is no longer a case of the instrument that serves the man, but of the man who serves the machine*. The novel characteristic of this interaction lies in that one of the elements in the relationship is no longer another human being, a master, but an entirely impersonal mechanism that Marx does not hesitate to qualify (rather romantically) as something that is "dead" as opposed to the living being incarnated by the worker.

As we know, Marx believed that this relationship depended not on the construction of machines as working means, but on the capitalist usage of them. He was convinced that, in the context of another social form of production, the relationship between man and machine could be reversed in favor of the former, and an ultimately humanized form of labor.

In actual fact, what Marx begins to foresee in some of its essential features is a new form of social life that develops under its own steam and irrespective of the economic model in which it first emerged. It stems not from any particular context, but from the structural form, functions and performativity of the leading characters involved. That the machine should be taken very seriously as a partner in a social relationship stems from the fact that its first appearance on the scene met with immediate, very strong negative reactions. Traditional workers saw the machines as formidable competitors and waged war against them right from the start. They initially imagined the destruction of this inanimate enemy as the only solution.

Such antagonism and the associated desire for revenge can only be explained by assuming that the workers implicitly acknowledged that the machines could really compete in the bid for work. This would not have happened if there were nothing human about machines, if they were wholly foreign to humans and their needs. But machines not only produce *for* humans, satisfying human demands, they are also the product of human expertise, they copy (repeatedly, and on a larger scale) certain human technical skills. It is in terms of these skills that machines can compete with humans, and they often win. The machine thus becomes one of the terms in a genuinely agonistic relationship, the possible outcome of which might be the destruction of one of the contenders, or the reciprocal improvement of both.

Marx grasps this situation very well, but he interprets it, once again, as the product of the essential confusion between tool and use. Machines rob the workers of their job because they become part of the capitalistic way of generating wealth. So Marx qualifies the struggle to destroy the new machines as "stupid", believing that the workers' attention should focus on the capitalist instead. If the machines crush the workers, if they override them and make the workers' skills so worthless that they may even lose their jobs, this is because the machines are being used by their owners to increase their profits. In another economic order, what Marx calls the mechanical means *in itself* would have very different effects:

> *The contradictions and antagonisms inseparable from the capitalist employment of machinery, do not exist, they say, since they do not arise out of machinery, as such, but out of its capitalist employment!* Since therefore machinery, considered alone, shortens the hours of labor, but, when in the service of capital, lengthens them; since in itself it lightens labor, but when employed by capital, heightens the intensity of labor; since in itself it is a victory of

> man over the forces of Nature, but in the hands of capital, makes man the slave of those forces; since in itself it increases the wealth of the producers, but in the hands of capital, makes them paupers [...].[8]

So the machine has in itself an essence entirely independent of the conditions in which it is employed. If the machine could operate outside the capitalistic production rationale, in an entirely neutral manner, operating on its own logic, it would *reduce* the working hours, it would *reduce* the intensity of the repetitions, and it would *free* the worker from the dominion of natural forces. But how can we configure this alternative usage more precisely? What idea of technological working does it draw on?

In actual fact, what Marx means when he imagines a non-capitalistic use of machines coincides with his idea of "humanized work", i.e. technological practices undertaken within the limits of the force available to the average human being. The machine *should work* as much as a man *can work*, and that is to say for a limited amount of time, at a rate that prevents him from becoming overtired, and with the general goal of releasing him from the constraints imposed by his work. Clearly, an approach of this kind can only stem from a decision that has nothing to do with the nature of machines, but derives from the "economic" rules that humans agree to apply to their use of these devices. Just as there can be a dehumanizing capitalistic way of using the machines that has the effect of crushing the workers, there can also be a "human" usage of the same devices. The method chosen depends not on the machine in itself, but on the boundary conditions imposed by human relationships. Marx sees the machine in itself as neither good nor bad, but it can be set to different – good or bad - economic uses.

It goes without saying that Marx also sees the devices manufactured by man as a sort of externalized *alter ego*. They emerge from eminently human needs and capabilities. But our relationship with our machines unequivocally triggers opportunities for us to make significant changes to the conditions that enable us to inhabit our world. This circumstance may be interpreted differently depending on the degree of variability that we attribute to so-called human "nature". If we see this nature as having a permanent, unchangeable "measure", our relationship with our machines will inevitably carry an intrinsic risk of dehumanization. On the other hand, if we think that humanity has not been cast in an inextensible "mold", and that man is a variable entity that can be shaped and adapted to the circumstances, then the new stimuli deriving from our symbiosis with our devices could lead to an alienation that will certainly change us, but may not necessarily be dehumanizing.

It is also to Marx that we owe some of the most profound analyses on how people's working conditions changed as a result of the introduction of machinery and mass production:

> In so far as machinery dispenses with muscular power, it becomes a means of employing laborers of slight muscular strength, and those whose bodily development is incomplete, but whose limbs are all the more supple. The labor of women and children was, therefore, the first thing sought for by capitalists who used machinery [...]

[8] Ivi, p. 464, p. 568.

> If machinery be the most powerful means for increasing the productiveness of labor — i.e., for shortening the working-time required in the production of a commodity, it becomes in the hands of capital the most powerful means, in those industries first invaded by it, for lengthening the working-day beyond all bounds set by human nature. [...] Hence that remarkable phenomenon in the history of modern industry, that machinery sweeps away every moral and natural restriction on the length of the working-day. Hence, too, the economic paradox, that the most powerful instrument for shortening labor-time, becomes the most unfailing means for placing every moment of the laborer's time and that of his family, at the disposal of the capitalist for the purpose of expanding the value of his capital.
>
> The immoderate lengthening of the working-day, produced by machinery in the hands of capital, leads to a reaction on the part of society, the very sources of whose life are menaced; and, thence, to a normal working-day whose length is fixed by law. Thenceforth a phenomenon, that we have already met with, namely, the intensification of labor, develops into great importance. [...] The shortening of the hours of labor creates, to begin with, the subjective conditions for the condensation of labor, by enabling the workman to exert more strength in a given time. So soon as that shortening becomes compulsory, machinery becomes in the hands of capital the objective means, systematically employed for squeezing out more labor in a given time. This is effected in two ways: by *increasing the speed of the machinery*, and by *giving the workman more machinery to tent.*[9]

The real relationship between man and machine involves specific actions, cognitive and motor patterns that are learned and then repeated with a variable frequency – and this is basically what happens for any type of human activity, not only for work.

Marx begins to see that the advent of the machine as a privileged partner leads to changes in some of the parameters governing the performance of human actions as part of the working process. The power mobilized by machines can be applied with a minimal effort by any human operator of average ability. This has the effect of making anybody a potential worker, not only adult males, but women and children too. Of course, we cannot fail to agree with Marx's condemnation of this situation, given the way in which this opportunity was exploited. But the end result is the concept that, in the world of machines, where human strength is no longer important, anyone can interact with these mechanical devices and use them to take action. Man interacts with the machine in a very natural way, and individual differences no longer significantly influence this interaction.

The time spent in action also changes dramatically. We can operate the machine without any limits on the working day. The machine becomes a sort of "lasting symbiont", a discrete, untiring company throughout a person's waking day. Our body needs even quite long periods of rest and sleep, but as soon as we are ready, we find the machine waiting for us and we take up from where we had left off. Slowly but surely, with the help of our machines, our living hours tend to be transformed into working hours.

The type of action required of the machine operator is no longer in any way comparable with that of a man handling his tools. The frequency with which the action on the production line can be repeated without the global quality of the process suffering leaves absolutely no space for the slow work of the craftsman. The

[9] Ivi, p. 414 e sgg., p. 517 sgg.

machine intensifies the human action, and this acceleration of the process has a feedback effect on the human operator, whose relationship with the machine induces him to develop mechanical movements that he also uses in his other actions and interactions with the world.

So far, all this has been seen as one of the ways in which human beings have been dehumanized. But nowadays we are not so sure that this is necessarily the case.

Chapter 3
Ernst Kapp – Organ Projection

Abstract In this chapter is discussed the important theory expressed by Kapp in the work *Grundlinien einer Philosophie der Technik* (1877). In his vision Kapp supports the idea that mankind was the engine principle of history and culture, who used technics to submit the world in the process to self-awareness. Kapp argued that everything mankind produces, as a result of projection of its organs, is to be interpreted as technics, then the only road to self-awareness leads through technics. Moreover, the only thing mankind can learn about itself is that in essence mankind is a technical species. This double movement is called "law of organic projection". In our interpretation he is true not only for the tools but also for each type of machine.

Keywords Kapp • Philosophy of technique • Organ projection • Tools • Steam engine • Telegraph • Culture • Selfconsciousness

Ernst Kapp's Elements of a Philosophy of Technology (*Grundlinien einer Philosophie der Technik*) were published in Braunschweig in 1877. Few people noticed the book, and even less attention was paid to the figure of its author – an elderly gentleman who had taken up a modest appointment as lecturer at the University of Düsseldorf on returning to Germany after a lengthy period of exile in the United States. In actual fact, Kapp was one of the most ingenious minds to contribute to the great German culture of *Naturphilosophie*, a follower of Humboldt and a thinker capable of further developing Hegel's thoughts in unexpected and original directions.[1] Considered a pioneer of the philosophy of technology, Kapp's name became attached to an approach to the nature of technological objects that is expressed very clearly right from the foreword of his *Grundlinien*:

> First of all, it is demonstrated by means of incontestable facts that man transfers (*uberträgt*) the shape, function and normal proportionality of parts of his body to his handiwork, and it

[1] On this issue, there is the fundamental study by B. Timmermans, *L'influence hégélienne sur la philosophie de la technique de la technique d'Ernst Kapp*, in Chabot-Hottois (ed.) *Les philosophes et la technique*, Paris Vrin 2003, pp. 95–108), which recalls the Hegelian derivation of Kapp's core concept, as well as its innovative potential, giving us a foretaste of ideas later expressed by Leroi Gourhan and Canguilhem. Already in the essay *Form und Technik* of 1930, Cassirer had traced a marked continuity on the topic of technology between Hegel, Marx and Kapp. It is also worth mentioning the very interesting pages that A. D'Atri dedicates to Kapp in her *Vita e Artificio. La filosofia di fronte a natura e tecnica*, Rizzoli, Milano 2008 (on pp. 165–172, in particular).

F. Grigenti, *Existence and Machine*, SpringerBriefs in Philosophy,
DOI 10.1007/978-3-319-45366-8_3

> is only afterwards that he becomes aware of these analogies. This fashioning (*Zustandekommen*) of mechanisms based on organic models, like the interpretation of the human body in terms of mechanical devices and, above all, the adoption of the principle of organ projection (*Organprojektion*) as the only way to achieve the goals of human activity, form the specific content of this volume (p. VI). (*our translation*)

As a geographer and scholar of primitive cultures, Kapp had more than once encountered the problem (well known to the paleoanthropologists) of how to explain the resemblance between the appearance of the objects and tools produced by man and the anatomical structure of parts of the human body. The similarities not only concern the shape, but also extend to the use made of the artefact, which very often reproduces the particular action that a human organ takes in the economy of the body's vital processes.

According to Kapp, this does not happen by chance. It is the result of a genuine transfer of *form* (*Gestalt*) and *function* (*Funktion*) from the body organs to the tools (*Handwerke*). This transposition is initially made unwittingly, however: tools may simply be found (a stick or a stone, for instance) or fashioned (an axe) without any clear awareness of the process underway. The first person to construct a hammer was not deliberately copying the shape of a human fist; he was simply replicating a shape that had proved effective, heedless of the anatomical structure from which it originated. Once an artefact has reached some degree of perfection, however, its performance enables us to gain a better understanding of the workings of the organ from which it originates. Men had been building mechanical pumps long before they knew that the heart serves the same purpose in the general economy of the human body; and it was from watching a pump at work that Harvey rightly guessed, for the first time, at the physiology of the cardiac muscle.

The projection of organic forms and functions is not a one-way process, from living to artificial objects. It also works in reverse, increasing at every stage of technological development what Kapp calls awareness of the self (*Selbst*), a term that he uses to indicate the human body, equipped with "life and flesh" (*Leben und Leib*), and its performance is the product of a combination of thinking processes and the particular structural mechanics expressed in the anatomical arrangement and capacity for movement of the body's parts, among which the hand has a very special role. It is the principal means for transferring the organic into the artificial body of objects, not only because it serves as a model on which the forms and functions of many tools are based, but also because the hand is the organ that governs and controls the use of all kinds of tools. An axe and a hammer are quite different utensils: the former recalls the shape of a human tooth, while the latter is reminiscent of a fist; but both are grasped, raised and operated by movements of a man's arm and hand. Much the same can be said of knives, etching tools, and any other utensils that can be made and used consistently with the human hand's capacity for movement and dexterity.

Kapp makes the point that, when we handle tools, we do not merely reproduce a function that a bodily organ is capable of performing, we also make its action more powerful. The primitive people's spears replicated the action of a man extending his arm to hit a target some distance away, but they succeeded in doing so incomparably better than the model organ was capable of achieving. This extension of the distance that could be covered gradually increased as technology devised ever more complex devices, and it played an essential part in the development of measurement and vision systems. The length of a human foot and the volume of a handful may be rather small quantities in themselves, but multiplied often enough they can be used to measure the world. The same is true of optical devices: magnifying glasses, binoculars and telescopes all reproduce human vision, but using them enables us to see things – large and small – well beyond the range of the human eye. Generally speaking, there are evident traces of an "anthropological scale" (*anthropologische Maasstabb*) embedded in the domain of the artificial that betray the eminently human (and consequently neither natural nor divine) origin of all those aids in which our species has unwittingly reproduced and extended particular capabilities of the human body.

As already mentioned, Kapp sees the principle of organ projection as belonging to the realms of the subconscious. When inventors design new devices they are unaware that they are creating something that reproduces or reflects a certain capacity of their own body in some way. They are focusing on other goals, to reduce the physical effort needed in some human activity, for instance, or to make a machine perform more cost-effectively. Even subsequent refinements of a new apparatus once it has been put to use do not seem to be guided by any conscious organ projection. Neither Watt nor Stephenson had in mind that connecting a steam engine to certain drive elements would lead to the construction of the locomotive and the expansion of the railways, and the wholly unpredictable end result was the development of a vast network of railway tracks that now covers the surface of the Earth like an immense circulatory system. Kapp constantly returns to such analogies, amply describing their effects, in the conviction that man-made tools and machines can unintentionally become the object of representations and judgements, and consequently instances of a greater self-awareness on a spiritual level too. Though rooted in the subconscious, the law of organ projection is not seen as a blind, instinctive tendency. Like all spiritual laws, it would ultimately give humans the opportunity to express their nature and see it reflected in their artefacts – in their tools and machines as well as in the great works of art and literature – that they produce for their own purposes. In the case of technology, humans not only generate strange analogies between the structural design of their artefacts and their own bodily organs, they also recognize themselves as living beings capable of technically humanizing their world.

Kapp acknowledges that it is easy to see the principle of organ projection at work when we look at the tools used by the craftsman or blacksmith. It becomes more difficult to explain how organ projection plays a part in the case of more complex machinery, the appearance and operation of which hardly seem to draw in any way on the human body as a model. Taking as an example the kinematic system consisting of an engine, drive elements and a machine tool, Kapp emphasizes the continuity existing between our hand tools and the tooling installed on machines: apart from their dimensions and other minor details, they are both artefacts that replicate the forms and functions of bodily organs. While the former are guided by a human hand throughout their working cycle, the latter do their job automatically, with no need for a human body to become involved. The huge power hammers that have been shaping gigantic sheets of steel since the dawn of the metalworking industry are still the organic projection of a human fist (just like the blacksmith's hammer), even though their "beating" action (their particular function) no longer takes the form of an arm-and-hand assembly, but has evolved into a fully-mechanized drive shaft.

Kapp attributes such developments to a purely technological reason, i.e. the introduction of the steam-driven machine in every sector of human activity. Designed and built by Watt without any precise idea of its potential uses, it quickly spread all over the world, soon becoming the "machine for all machines". Reflecting the adaptability of the human hand (which is capable of wielding all kinds of tools), the steam engine can be connected to all kinds of machinery. Its fundamental distinctive feature lies in its universal versatility. Its second characterizing feature concerns the relationship that this new apparatus establishes with the various sources of energy: the steam engine puts the planet's natural elements – earth, water, wind and fire – to a new type of use, obtaining a vast, uninterrupted driving force. Bearing these particular features in mind, Kapp wonders in what way, and in which of its features, the living body has served as a model for the construction of such machines. How does the principle of organ projection apply to the case of the steam engine?

As Kapp sees it, the problem seems particularly complex if we bear in mind that certain characteristics of the living body cannot be brought down to mere automated actions – and the most important of these concerns our goal-oriented behavior. The apparent body movements of all sensitive animate beings are constantly being adjusted in response to changes in the state of their objectives. When a cat chases a mouse, it adapts its running action and displacements to those of its prey. If the mouse suddenly changes direction, the cat instantly follows suit; if the mouse finds somewhere to hide, the cat will crouch and wait for it to reemerge; and so on. In animals, intention and volition work as an additional non-mechanical control system capable of inducing a change of state in the kinematics of the body system. Kapp strongly emphasizes this aspect:

> The machine is an artefact built by an external will, while the human body grows *ex ovo* according to an inherent, hidden law. The *egemonicon* of the machine is not intrinsic, nor does it belong to the machine; the stoker of the steam engine and the driver of the locomotive govern their machines just like a jockey on his horse. Instead, the *egemonicon* of living beings (their will and intelligence) are intrinsic, a constitutive and integral part of them. Drawing an abstraction from their physical functions, we could say that the parts of a machine always remain identical to themselves until such time as the machine has to be repaired, whereas the parts of a living being only remain the same in terms of their shape, while their substance changes continually, it is self-regenerating and self-repairing (Kapp, p. 132). (*our translation*)

The machine and the body have different origins: while the former owes its creation to the will of a human designer and manufacturer, the living being develops on its own, based on inner laws that govern a matter capable of self-organization right from the start. No mechanism can be self-generating and self-governing: its very existence and purposes are entirely reliant on human beings. The organic body, on the other hand, comes into being and grows, already containing within itself the principle of its own existence and its every subsequent implementation. According to Kapp, volition and reason are none other than the highest spiritual expressions of this living body's autonomy. It is this structural difference between man and machine that explains why, once assembled, the physical parts comprising a machine undergo no change; they remain exactly the same from every point of view up until they may need to be replaced, whereas living matter remains stable only in form, while it grows and regenerates itself with no outside intervention.

There is therefore – even in terms of exterior appearances – a huge difference between mechanical devices and organic bodies. Whereas tools may look more or less like faithful replicas of body parts, machines can have a shape and size on an incommensurately larger scale than those identifiable in the organic world. Kapp claims, however, that machines are perfectly capable of performing actions (*Leistungfähigkeiten*) typical of the organic body even when they bear no resemblance to a human being at all. When considered in terms of intensity of repetition and quality of execution, moreover, their performance is all the more remarkable, the more their physiognomy differs from that of a human being. This is exactly the case of the steam engine:

> What inspires our utmost admiration for the steam engine lies not in the single technical details – like the reproduction of an organic connection of members by means of planes revolving on lubricated surfaces, for instance, or the bolts, connecting rods, firing pins, levers, and pistons – but in how the machine is powered, the transformation of the fuel into heat and motion or, in other words, the devilishly particular appearance of its own autonomous capacity for work. (Kapp. p. 138). (*our translation*)

Unlike what happens with tools, machines reproduce not functions related to the form of a given organ, but more complex processes. In this sense, Kapp considers the case of the steam engine emblematic. It would be pointless to seek evidence of the principle of organ projection in action in its exterior appearance; it does not replicate organic body parts or reflect anatomical details. But it does recreate the living body's capacity to convert energy (contained in foodstuffs) into mechanical work. This capability (impossible for any manual tool) enables artificial entities to move unassisted for the first time, without any external driving force, be it man, wind or water. When Watt's machine was connected to a pump, a rolling mill or a weaving loom, and eventually to a wheel-mounted cart, it soon became clear that the capacity for *animation*, or the principle that makes organic beings capable of unassisted motion, had been transferred once and for all to the field of machines.

Another example of the projection of "higher" organic functions concerns the electric telegraph, to which Kapp dedicates a very ample chapter in his book. His reference model in the organic world is the nervous system, that Kapp sees projected in the core element of the new technology: the electric cable replicates not only the function (the capacity to transmit an impulse), but even the form of human nerve fibers. The principle of projection works very effectively in this case – to such a degree that the telegraph can be seen as humanity's nervous system and it is equally plausible to consider the human nervous system like the body's network of electric wiring. This perfect homology has a highly significant, paradoxical aspect:

here again, the functional transfer of capacities from the organic model to the technological copy was wholly unintentional, and this is all the more curious when we consider that Kapp does not hesitate to define the telegraph as "the machine of the spirit":

> The technology for manufacturing machinery reached a peak in its development when the steam engine, in terms of the concept of the storage of power on the one hand, and the development of the locomotive on the other, came to reproduce the body's vitality. The same happened with the telegraph in terms of the functions for the communication and transmission of thought, and again in that ultimate purification of raw matter called the "universal telegraphic", which expresses the greatest proximity to the forms of the Spirit (Kapp, p. 153). (*our translation*)

In conclusion, man reproduces in his world of machines not just single human organs, but his whole body too. Living movement and thought are not mere parts or elements separated from a whole, but general functions on which the system's overall organization depends. The principle behind the combination of different devices to enable the assembly of different machines makes it possible to build complex automata that are increasingly capable of eminently human activities.

The stance taken by Kapp has been variously interpreted. Seen from the side of the machine, it could be defined as an "organicist functionalism". In fact, Kapp does not claim that the organic projection of forms and functions makes machines indistinguishable from human beings. He sees as degrading the view taken by Helmholtz that the physical concept of 'work' coincides exactly with the type of activity that humans describe with this term. If this were true, Kapp suggests, if mechanical work were really indistinguishable from that of human beings, then humans could be replaced by machines in every kind of activity, including the design and construction of the machines themselves! But Kapp judges this outcome absurd and impossible. The human body's capabilities will be simulated better and better by machines, but never to perfection, and the total replaceability of humankind by machinery can therefore be imagined, but never actually put into practice.

This brings us to the core meaning of the principle of organ projection that Kapp expresses in a passage afforded little attention from the critics:

> The theory of organic development coincides with a practice of continuous mechanical improvement that has led from the first stone axes, through a variety of tools, apparatuses and devices, right up to that complicated mechanism that is the idea of the "model machine" [...] *intended as a type of physical apparatus that should serve for the purpose of understanding and reproducing the reciprocal action existing between physical forces and vital bodily processes* (cit. p. 133). (*our translation*)

The ideal machine – or rather the ideal of the machine, the device that all improvements made to technological means strive to achieve – is none other than the organic machine, the most advanced and effective product of Nature. It is in the living body that the energy conversion process essential to the system's survival is perfectly integrated with the control exerted by the higher spiritual functions. In Kapp's time, it was impossible to design or implement the transfer of such an assembly to the artificial world. His principle of organ projection thus imposes a very clear direction for any technological development: the purpose and ultimate goal of the artificial is not to replace the organic, but to become organic too. In other words, the living being is a perfect machine, and its capabilities delineate the ultimate limit that every device tries to equal.

The epistemological consequences of this approach are clear: on the one hand, our machines make us increasingly aware of our own organic endowment; on the other, we can already predict that we will never be able to improve our machines beyond the maximum capabilities (that we still do not know) of our human body. Simply put, machines do not evolve from machines; if anything, they represent an artificial development of the physiology of the living body. It will therefore be from advances in our knowledge in this area that we might expect a further growth in our ability to invent and construct artificial apparatuses that can only be copies – increasingly efficient, of course, but still copies – of the model represented by the living being.

Chapter 4
Ernst Jünger and Friedrich Georg Jünger – Dominion and Machine

Abstract In this chapter are confronted the different visions of the machine in the work of Ernst and Friedrich Georg Jünger. By Ernst Jünger we analyzes: (a) the concept of total mobilization (Die Totale Mobilmachung). With this notion Jünger means that real process which, thanks to the development of technology, makes it feasible in practice to mobilize the totality of the power (*Macht*) available in the universe. For Jünger, power is not the accumulation of a substance waiting to be developed and completed, but the rapidity with which energy is converted; (b) the idea of *perfection* of technology, the ultimate expansion of total mobilization, the final settling of the technological dynamism in the spatial domain, will be complete when that immovable point of balance is achieved where the machine has completed its conversion into an organic form, i.e. when the human body, in the *Typus of the Worker*, will have definitively acquired the structure of the mechanism. By Friedrich Georg Jünger is the distinctive way in which the machine mobilizes the world can be elucidated more clearly by looking at the category of *over-exploitation* (*Raubbau*). Friedrich Georg uses this term in the sense that it acquired in the German language in the technical vocabulary of the timber industry: *Raubbau* precisely indicates the practice of cutting down trees without reintegrating the stock to make up for the plants being destroyed.

Keywords Jünger Ernst • Jünger Friedrich Georg • Total mobilization • Power • Worker • Organism • Machine • Over-exploitation • Consumption

There is a clearly evident continuity between the works published by Ernst Jünger in the 1920s and 1930s – *Im Stahlgewitter* (1920), *Die Totale Mobilmachung* (1930), *Der Arbeiter* (1932), and the later works by his brother Friedrich Georg – *Die Perfektion der Technik* and *Maschine und Eigentum* (which both appeared in 1946, but had been completed in 1939). In fact, they could be seen as chapters of the same book or parts of the same theory. But apart from the fact that they were brothers (which certainly played a part), the two authors' shared reflections on the topic of technology have very little in common with other "duets" published in the recent history of culture. Both men lucidly pinpoint the timing of the passage to the dominion of the machine, but while Ernst (the elder brother) pauses with some satisfaction to describe the early and still uncertain stages of this passage, his younger brother

F. Grigenti, *Existence and Machine*, SpringerBriefs in Philosophy,
DOI 10.1007/978-3-319-45366-8_4

Friedrich Georg, already at the end of the 1930s, produces a scathing account of the complete and definitive *perfecting* of our technological form of existence.

As we know, Ernst Jünger analyzes the essence of technology starting from its definition as *the mobilization of the world implemented by the form of the Worker.*[1] It is worth stopping briefly to consider the single terms of this complex concept. By *total mobilization* (*Die Totale Mobilmachung*), a notion drawn by no means casually from the language of war, Jünger means that real process that, thanks to the development of technology, makes it feasible in practice to *mobilize* the *totality* of the *power* (*Macht*) available in the universe. This begins to happen in the modern world, and promptly takes on the nature of a necessity: though a historical fact, total mobilization is not a contingent factor, something that might have been avoided, but an inescapable appointment with their destiny for humanity as a whole.

The concept of *power* (*Macht*) introduced by Jünger is mediated by Nietzsche, but Ernst lends it a particular connotation deriving from a careful consideration of the real behavior of devices and materials. For Jünger, power is not the accumulation of a substance waiting to be developed and completed, but the rapidity with which work is done. In other words, it is the measure of the amount of work that can be done per unit of time and, since the work done in a given process corresponds to the energy converted in the process, we can conclude that power represents the *rapidity with which energy is converted.*

What transpires from the author's extraordinary descriptions of total mobilization is the idea that this entails a tumultuous upheaval of every order of existence, from the finest structures of matter to the ethereal constructs of human spirituality:

> The type of movement we are discussing, however, dominates not only the working rhythms of – cold or burning – artificial brains that man has created for himself, where the radiance of icy lights phosphoresces. It is a movement that is perceivable as far as the eye can reach and in this age our eye can see far. Secondly, the movement has not only got its hands on traffic –the mechanical overcoming of distance that aspires to equal the speed of a bullet – but on each activity per se. We can see it in the fields where we sow or we reap, inside the mineshafts from which iron and coal are extracted, at the dams which stop the water from rivers and lakes. It is at work in thousands and thousands of variations, from the smallest workbench to the large production sectors. It is not absent from scientific laboratories or from commercial agencies or from any public or private building. It is present both where we act and think, where we fight or where we have fun. Those who examine this language raise the issue of its essence. The simple and immediate answer is that this essence is certainly to be found in mechanicity. But, as the observational material accumulates, we are forced to recognize that in this context the ancient distinction between mechanical forces and organic forces has been overcome.[2]

[1] See above all the chapters from 44 to 58 in *Der Arbeiter*, in E. JÜNGER, *Sämtliche Werke*, Zweite Abteilung, Band 8, Klett-Cotta, Stuttgart 1981. On the matter of technology, also in connection with Heidegger's thinking, see: P. NERHOT, *Ernst Jünger-Martin Heidegger: il senso del limite (o la questione della tecnica),* Cedam, Padova 2008; as concerns the brothers' relationship, see above all F. STRACK, *Titan Technik. Ernst und Friedrich Georg Jünger über das technische Zeitalter*, Königshausen und Neumann, Würzburg 2000; and on technology and war: M. MÄNGEL, *Das Wissen des Kriegers oder Der Magische Operateur*, Xemonos, Berlin 2005.

[2] E. JÜNGER, *Der Arbeiter*, cit., p. 77 (our translation).

Jünger's diagnosis is clear: the only type of movement capable of putting every atom of matter-energy contained in the universe to work is mechanical movement. Not only can it be applied to bodies that have mass, it also models and reactivates all other movements, including that of human thought. Total mobilization can therefore only take place thanks to the *mechanicality* and particular characteristics of such movements, and primarily the acceleration that they induce once they have been applied to different human actions. From this perspective, the mechanism's movement ceases to be a partial activation reserved only for certain sectors of existence, becoming – in the era of the machine – the intrinsic quality of every possible form of dynamism. Every traditionally accepted distinction between different realms of activity must therefore be overcome. It is not only that organic forces are no longer distinguishable from mechanical forces, but even labor, political action and human thought can no longer be imagined as areas free from mechanical activation. We may well continue to act and think, but we shall do so in the manner permitted by the inexhaustible, accelerated and totalizing dynamism that the era of technology expresses as its essential quality.

According to Jünger, the first complete deployment of total mobilization occurs with the outbreak of the "battle of materials" characterizing the final stages of the First World War. The fighting between human beings is replaced by the massive deluge of fire and metal brought about by the warring nations in their final effort to prevail in a war that, after it ended, was to leave the winners in exactly the same state of prostration as the losers.

But the machine had already shown its sinister side during the very first infantry charges along the front lines of Europe. With cold precision, Ernst Jünger analyzes the massacre of Langemarck, where thousands of young German volunteers lost their lives, cut down by the latest-generation British machine guns. He describes the abrupt felling of a whole order of the spirit by the triggering of a repetitive mechanism capable of producing dead bodies on an industrial scale:

> In this episode we see the breaking of a classic attack in spite of the power present in the desire for strength that animates individuals and in spite of the moral and spiritual values that make individuals distinctive. Free will, education, enthusiasm, drunken disdain for death are not enough to overcome the sapping strength of a few hundred metres in which mechanical death wraps us in its spell. In this way, an extraordinary and really spectral image of dying in the realm of pure ideas emerges, a decline where, like in a nightmare, even an absolute effort of will cannot defeat a diabolical contrary impulse.[3]

The sudden arrival of the machine in this area of human action instantly undermined the long-established traditional hierarchy between the power of the ideal and the efficacy of the mechanism.

The "idea" – as the product of a specific human activity undertaken primarily in the arts, philosophy and literature – not only took pride of place in the government of human faculties and social classes, it also infused every other activity that might be conducted by an individual or by a community as a whole. Even war had to be tinged with the color of the highest values, such as a desire for justice, dedication to

[3] Ivi, p. 85 (our translation).

a cause, and courage. Only a heartfelt belief in such ideals and a strong determination to defend them could succeed and defeat not only the enemy (who was often animated by the same disposition of the spirit), but also the elementary forces taking shape in the instruments of war. A war was meant to be a human confrontation that took place for human reasons, and the use of machines was only a means serving this purpose. So the side with the stronger spiritual motivation should have been victorious irrespective of the power of the means employed.

This idea was amply disproved during the first charges of the infantry on the various fronts of the Great War: by spraying the field with thousands of bullets, i.e. by increasing beyond measure the mechanical frequency of the shots being fired, the machine gun came to take part in the ancient contest between the philosopher and the blacksmith, tilting the balance the definitively in favor of the latter:

> The deep core of the carnage/events that occurred in Langemarks founded on the onset of a cosmic conflict that always recurs when the universal order is shocked and that, in this circumstance, transforms itself into the symbol of a technological era. It is the conflict between solar fire and telluric fire, which on the one hand shines like a spiritual flame and on the other shines like a terrestrial flame – like light or fire. An exchange of spells between "the chanters to the barrow of sacrifice" and blacksmiths who tap the energies of the metals of gold and iron. The bearers of the idea that, moving away from the archetypes has become a most beautiful copy, are knocked down to the ground by matter, the mother of things.[4]

The machine emerges on the invisible boundary that marks the threshold of a new order of the world. The sunlight that once made the idea visible makes way for the flickering fires of the workshop. Another scenario is coming into view and our gaze shifts from archetypal models now showing signs of wear to a space where matter has returned to its primitive state of pure receptacle, waiting to be processed by machines. The latter emerge from the trenches of wartime and move on to populate the immense technological space opening up in the suburbs of the great industrial cities, becoming ever more powerful as they do so.

In this transition, the machine acquires a more and more essential, simplified form. Within the space of a few generations, the early complex devices (which were difficult to use and often likely to malfunction) are replaced by machines designed with bodies comprising fewer and fewer components. Their dimensions change and their shape becomes aesthetically recognizable, almost as if we were seeing the evolution of an organic species. The machine changes; the initial, chaotic multiplicity of shapes becomes consolidated in a definitive state of perfection that makes any further development impossible.

In the beginning, the machine occupied the traditional workplaces, the domestic workshop and the factory, substituting the human hand in certain, clearly defined processes. For a long time, it remained confined within a limited space, foreign to our living social world. It was not until the start of the twentieth century, after it had left the battlefield, that the machine began to invade and change every aspect of human life:

---

[4] Ivi, p. 86 (our translation).

> Many examples could be given but it is sufficient to watch the theatre of our lives in its exuberant unfolding and relentless discipline, with its smoking production areas and twinkling of lights, with the physics and metaphysics of its trade, its engines, airplanes and metropolis teeming with people, in order to apprehend with a sense of dismay and excitement that here there is not one atom that is not working, and that this delirious process is, at its core, our destiny. The total mobilization occurs on its own, it is, in war as in peace, the expression of the mysterious and inexorable law that gives us the age of the masses and of machines. Thus it happens that every single life always edges without discussion to the conditions enjoyed by workers, and that the wars of knights, kings and citizens are followed by the wars of workers, wars whose rational structure and implacability has already been suggested by the first great conflict of twentieth century.[5]

Passages like these give us a better idea of Ernst Jünger's fundamental theory. The purpose of the machine – in the sense of a technological means-environment, that expands by itself with no precise intentionality – is not progress or profit, but *dominion.* First of all, it destroys a certain political order with a view to establishing a new, even more ruthless legislation. The first action that machines take is to demolish all the traditional hierarchies, along with the principles that supported them. King, knights and priests are obliged to make way for a new type of human being, the Worker, whose primary *raison d'être* stems from his direct relationship with the machine.

In addition to destroying the previous order, every new dominion immediately establishes new, alternative forms of government and criteria for selecting the new élite that is to take the lead in the management of the public space. As technology expands, mobilizing the totality of the world's resources, the space left for nurturing non-technological activities shrinks, becoming smaller and smaller. Even politics, religion and all the traditional forms of human culture very soon take on the connotations of work that can be done by means of machines. Working thus ceases to be just one of an individual's possible activities, it becomes the determinant that characterizes all social practices. We are working whatever else we are doing, even in our spare time, whenever and wherever we engage in even the most insignificant movement. What Ernst Jünger calls *total work* is neither a new form of work that is added alongside existing forms, nor the old form of the artisan's work that suddenly takes control; it is a new type of exercise to which all others must conform.

The figure of the Worker must consequently not be seen as representing a given social class or an individual employed in a given activity. The Worker incarnates a new *form of existence* with connotations that are more ethical and metaphysical than in the manner of a historical or social contingency. Once the technological environment has invaded the whole space of human affairs, criteria are established for selecting individuals particularly capable of managing the new means. Machines demand an adequate training, and this makes it necessary to set new, more efficient levels of human performance. For their part, individuals having to express an opinion on technology are faced with precisely two options: either they can accept the

---

[5] E. JÜNGER, *Die Totale Mobilmachung*, in *Sämtliche Werke*, Zweite Abteilung, Band 7, Klen-Cotta Stuttgart 2002, p. 357 (our translation).

new life form and seek to dominate its movement; or they can refuse it, but in this latter case they will very soon be cast aside:

> Wherever man falls under the jurisdiction of the technique, he sees himself placed before an invisible ultimatum. There is nothing left to him but a choice: either to accept the tools of technique and to speak its language, or to sink. But if he prefer to accept, and this is very important, he becomes not only the subject of the technical processes, but, together, their object. The use of instruments implies a certain lifestyle, which relates to big and small circumstances of life. The technique is therefore not a neutral force, not a reservoir of effective or convenient means by which any of the traditional forces can draw on its discretion. Just behind this appearance of neutrality lies rather the mysterious and seductive logic with which the technique is ready to be at the service of men. This logic is becoming more obvious and irresistible in proportion to the impulse with the space where the work earns totality. In the same proportion it is becoming weaker instinct of those who are struck by the technique. Instinct was that of the church, when it wanted to destroy a knowledge which saw a earth as a satellite of the sun; instinct was the knight who disdained rifles, or weaver who destroyed the machines, or the Chinese who prevented their entry into his country. But they all signed their peace treaty: the kind of peace that reveals defeat.[6]

Technology radically modifies the identity of any user. It is a mistake to imagine technology like a huge box of tools that traditional man can open, exploit, and then hope to close again and walk away unscathed. Every technological means imposes movements, behavioral rules, particular exercises that very soon come to be seen as the only way of taking action and of coming about. The effort involved modifies a performative balance that had traditionally been set as the level to achieve. The dynamism existing in the world of machines does not only generate higher performance limits, it also imposes a discipline based on the opposite principle, i.e. no limit is really insuperable.

Jünger believes that making a deliberate decision to actively adopt the new lifestyle can be the first step in an effort to impose a renewed rehumanization of the technological space. Once the anarchism typical of the era marking the passage to technology has given rise to a world order wholly dominated by the machine, the form of the Worker will have reached the same degree of perfection as the machine. He will therefore be able to deal with the machine in a relationship of mutual parity, as happens in a space of forms that, after engaging in a lengthy battle, are not destroyed, but undergo a selection process.

Jünger makes a characteristic use of the term *form* (*Gestalt*): he means not an essence, but rather the maximum degree to which it can be manifest. Carburetors and direct injection systems are two different structural solutions for supplying power to the same form of combustion engine; whichever solution is adopted, the means of propulsion will continue to function as a result of internal combustion. We might debate which of the two operating modalities best expresses the essence of the technological type known as the "internal combustion engine". According to Jünger, every machine (like the whole fact of technology) possesses a maximum degree of irradiation, which corresponds to its perfection (*Perfektion*). Having

[6] E. JÜNGER, *Der Arbeiter*, cit., p. 134 (our translation).

reached this stage of its development, all innovative dynamism should cease, the forms become stable, and no further improvement is possible:

> In reference to these tools, it is possible to speak of organic construction at the time the technique reaches that highest degree of naturalness that is present in the limbs of the animals and in the joints of the plants. Even in the embryonic state of technique, such as that in which we find ourselves, we cannot neglect the existence of an effort directed towards not only improved profitability, but also to the effectiveness with a more daring simplicity of lines. We are testing how this affair cause a more lively satisfaction not only to the intellect, but also to eye-and causes it with that lack of intentionality that is one of the characters of organic growth.[7]

The utmost manifestation of the machine form thus corresponds to the achievement of organic perfection. Once artificial devices have reached the level of structural efficiency detectable in the anatomical structures of animals and plants, the tumultuous evolution in their development will stop because it will have reached its limit. From this moment onwards, no further improvement will be possible. Technological forms that have become established in their maximal expressivity, like organic bodies, will acquire an ideal value and serve as perfect, eternal models. Incessant movement will be replaced by stasis. A space where every point was mobilized, pervaded by lines of force of unpredictable intensity, will be replaced by the invariability of power diagrams. Jünger believes that it is only at this stage that the measure of control that the Worker will have succeeded in exerting over the machines can be accurately assessed. But, in the meantime, on the grounds of what specific perfection can he freely dispose of the technological space?

Jünger's answer emerges from various comments dotted here and there amidst his pages: doubts about an incompliant future make the seer cautious, obliging him to sketch with a light hand an image that continues to appear unclear and indefinite:

> In this area, images of a supreme discipline of heart and nerve worthy of a corollary of the improvements traditions, have become history: evidence of supreme, unadorned, almost metallic coldness, from which a heroic conscience can manipulate the body as a pure instrument and tear, beyond the limits of instinct of self preservation, still a series of complicated performances.[8]

Even in the most advanced stages of technological development, the Worker has yet to definitively sever the link with the elemental resources of his existence, which are courage in the face of danger, amorous passion, emotion when exposed to risk. He can still use these forces as tools to fashion his own individuality. The Worker mobilizes the world, but he also mobilizes himself in view of the necessity, imposed by technology, to reach that degree of perfection – among those achievable by human beings – that is best suited to the dynamic movement operating in the space dominated by total work. But with what *habitus* does this human capability coincide?

---

[7] Ivi, p. 142 (our translation).

[8] Ivi, p. 86 (our translation).

As mentioned in the last short quotation, this should consist in the stable adoption of the capacity to use the body as a *purely mechanical instrument* or, in a nutshell: *the most perfect stage of the Worker appears when the form of the organism has been transformed into that of the machine.*

This stepping over the line between organic and inorganic does not happen suddenly, but only after a lengthy period of exercise governed by a very strict discipline, the purpose of which is to improve performance with a view to achieving a new, higher level of efficiency. Once it has been reached, this new *record* becomes a reference level and all further exercises strive to surpass the new limit.

The *perfection* of technology, the utmost expansion of Total Mobilization, the final settling of the technological dynamism in the spatial domain will be complete when that immovable point of balance is achieved where the machine has completed its conversion into an organic form, i.e. when the human body, in the type of the Worker, will have definitively acquired the inflexible structure of the mechanism.

Friedrich Georg Jünger returns to the topic of the machine along the lines already charted by his brother. Unlike Ernst, however, he does not believe that the state of perfection that will be reached by technological dominion will be in the sense of a balance between the totalizing organization of the machines and human needs.

Starting from the idea that every act of rationalization of the world stems from the awareness of a "shortcoming" that humans see as a weakness to be overcome, Friedrich Georg interprets the imperious development of technology as nothing more than the extreme expansion of *poverty* (*Armut*), that he sees as the essential determinant of man's condition. The goal of technological dominion would be not to increase wealth, but to keep humans in that congenital state of irremediable shortage of resources that provides the very foundations of technological development. Through the mechanization of work, and every other aspect of life, pauperism simply extends its power unchallenged, even in times when the techno-economic system seems to be enjoying a temporary phase of enrichment.

On this general view, Friedrich Georg grafts some more specific considerations on mechanical devices. In *Die Perfektion der Technik (The Failure of Technology)*, he admits that we have to acknowledge machines the most primitive form of human intelligence, which is a faculty of assemblies and structures that enables even an artificial device to process elements of nature in a manner enabled by applying physical forces. Beating, pressing, or forging are operations that machines can handle precisely and more productively than human beings. But a careful look at what happens in a factory when the machines are in action will reveal that, behind their enormous, incessant movements, other goals are being pursued that are very different from those appearing on the surface of the automatism:

> The impression we gain as we observe technical processes of any sort is not at all one of abundance. The sight of abundance and plenty gives us joy: they are the signs of a fruitfulness which we revere as a life-giving force. Rooting, sprouting, budding, blooming, ripening and fruition – the exuberance of the motions and forms of life – strengthen and refresh us. The human body and the human mind possess this power of bestowing strength. Both man and woman have it. But the machine organization gives nothing – it organizes need.

> The prospect of vineyard, orchard or a blossoming landscape cheers us, not because these things yield profits, but because of the sensation of fertility, abundance, and gratuitous riches. The industrial scene, however, has lost its fruitfulness; it has become the scene of mechanical production. It conveys, above all, a sense of hungriness, particularly in the industrial cities which, in the metaphorical language of technological progress, are the homes of a flourishing industry. The machine gives a hungry impression. And this sensation of a growing, gnawing hunger, a hunger that becomes unbearable, emanates from everything in our entire technical arsenal.[9]

As we can see, the author clarifies the essence of mechanical processes by comparing two orders of productivity. In one, there is nature's exuberant capacity for generation that expresses itself most beautifully in the phenomena relating to birth and growth. In the other, we find mechanized industry, with its remarkable productive dynamism dominated by the need to satisfy hunger – intended here in the sense of a sort of metaphysical need capable of shaping itself in unpredictable ways. While we find excess and pointless expenditure on the one hand (it is no chance that Friedrich Georg recalls the phenomenon of the gift), on the other the prevailing logic aims to maintain indefinitely a state of empty penury. The apparent abundance of resources set in motion by machines conceals the sense of a process that is dedicated entirely to consumption and destruction.

This view is modulated along different lines, which can be summarized in the following terms.

Not only machinery, but also the rational order behind it, is revealed as a manifestation of hunger; the exponential growth in consumption (*Verzehr*) is an effect not of overabundance, but of poverty and the associated phenomena of concern about the future, need, and an increasing work-related fatigue.

We should not expect an extension of the technological form of production to provide any real solution to the problem of how to satisfy man's primary needs; technology is the very expression of these needs, and every solution that it might excogitate simply amplifies the extent of our penury and emptiness.

The distinctive way in which the machine mobilizes the world can be elucidated more clearly by looking at the category of *over-exploitation* (*Raubbau*). Friedrich Georg uses this term in the sense that it acquired in the German language in the technical vocabulary of the timber industry: *Raubbau* precisely indicates the practice of cutting down trees without reintegrating the stock to make up for the plants being destroyed. Wood obtained in this way will certainly serve to make furniture and other products, but once they too have been used up, there will remain *nothing* of the original substance.

To anyone objecting that such a model is due not so much to the machine as to the economic use that is made of it, Friedrich Georg responds:

> In every healthy economy, the substance with which it works is preserved and used sparingly, so that consumption and destruction do not overstep the limit beyond which the substance itself would be endangered or destroyed. [...] Since technology presupposes

[9] F.G. JÜNGER, *Die Perfektion der Technik*, Klostermann, Frankfurt a. M. 2010, p. 27 (Eng. trans., by F.D. Wieck, *The failure of Technology*, introd. by F. D. Wilhelmsen, Chicago, Gateway Editions, Chicago 1956, p. 19.

destruction, since its development depends upon destruction, it cannot be fitted into any healthy economic system; one cannot look at it from an economic point of view.[10]

The idea that profit-seeking is the real reason for the indiscriminate exploitation of the resources needs to be overcome. An economy of pure consumption exists simply because a good administration has incorporated the use of machines, which have subsequently perverted their intended use. The wise use of primary elements, the sustainability of the levels of consumption, and the reintegration (even only partial) of what is destroyed are overridden by the annihilating appetite of the machines. As a result of their movements, the substance of the world is undergoing a gradual, unstoppable process of destruction. Nothing can stop the appetite of the machines. No richness is large enough to escape them. No rational law is capable of imposing a limit on them. The perfection of the machine coincides with the total consumption of the world.

Where exploitation arrives, there begins the devastation (*Verwüstung*). It takes concrete shape in the increasingly common image of the industrial city, the appearance of which depends on the type of technology dominating therein. Friedrich Georg sees Manchester as representing perfectly, in its ugliness, the monstrosity of the steam engine. So there is an aesthetic of the machine that does not stop at the object expressing it, but also spreads like a disease to its surroundings and to the inhabitants, who tend increasingly to resemble the machines with which they deal day to day. The term *Häßlichkeit* used by the younger Jünger to indicate the distinctive aesthetic trait of this devastation, means not only "ugliness", but also "badness" and "wickedness". The original value of beauty is converted into its exact opposite; and where something of beauty still survives (especially in places where the machine has yet to arrive), special action must be taken to defend it so that what little of substance remains is not definitively depauperated too.

The machine's hunger does not spare human beings. The proletarization of work, reducing thousands of men and women to a condition of exploitable resource, finds its ideological expression in the concept of a working occupation as a right, and the acknowledgement of this right as what characterizes the individual. Friedrich Georg makes the point that everyone converges on this idea, from right and left, while people's lives tend to be increasingly phagocytosed and transformed by the movement of machines. Even when the amount of time actually spent on the job is shorter, people spend more time interacting with machines that entertain them in their so-called "free" time.

Finally we come to politics. According to Friedrich Georg Jünger, proof of the incipient achievement of a state of perfection by technology can be seen in the insistence with which, already in his time, people called for a "technical government". He had already perceptively pointed out that, when economic recessions put politics (and even economics) to the test, people begin to voice the idea that a "technological planning" of the political decisions would generate more effective solutions when

[10] Ivi, p. 20.

the traditional government shows clear signs of inefficiency and "technical" unpreparedness. In the younger Jünger's words:

> When economic crises can no longer be overcome by economic means, human hopes turn toward a stricter rationalization of technology: the idea of technocracy arises. But first we should examine whether it is not technology itself which brings about such crises. We should examine whether technology is capable of putting our economy in order and whether such an ordering falls within the scope of its tasks at all. What does "technocracy" mean? If the word has any meaning, it can only be that the technician rules, that he takes over government. But the technician is no statesman; he has no talent for politics. His knowledge is one of technical, functional effects. All technical knowledge is marked by an impersonalism that necessarily results from the purely material facts that it deals with. This impersonalism is reason enough to doubt whether the technician is capable of taking over and running the affairs of state.[11]

Technology shows that it has reached a state of perfection when it proposes in the first person to govern the state. It sees the "personalism of politics", with its passion and partiality, as obsolete. The idea, we hear say, is to do away with politicians and replace them with technicians who, instead of adopting the criterion of justice, will deploy objective knowledge of the laws governing the various machines: the state, the economy, and society.

Friedrich Georg Jünger's diagnosis leaves no space for hope – and we now know that he foresaw what is before everyone's eyes today.

[11] Ivi, p. 26.

Chapter 5
Martin Heidegger – Machine and Truth

Abstract In this chapter are interpreted certain passages of Heidegger's essay *Die Frage nach der Technik*. Special attention is given to the concept of *enframing* (*Gestell*). In our reading this term refer to mechanical supporting structures, i.e. to devices that are not particularly complex, but they serve fundamental purposes such as supporting or containing in the broadest possible sense, so we might also translate *Gestell* as "assembly" or "framework". With Gestell Heidegger attempts to express the essence of modern technology in the way that it reveals its "mechanicality". The assembly is not just a device among others, but the horizon on which all the resources of the technological world, including human beings, are collected and coordinated. The object touched by the machine is thus converted into a "standing-reserve" (*Bestand*), or stock, something to be extracted, converted and consumed.

Keywords Heidegger • Gestell (enframing) • Technik • Poiesis • Bestrand (stock)

Heidegger remained profoundly affected by Ernst Jünger's works. He read *Total Mobilization* (1930) and *The Worker* (1932) immediately after they were published and repeatedly discussed them with colleagues and friends up until the winter of 1939–1940, during which he also read Ernst Jünger's short story *On the Marble Cliffs* (published in 1939). Heidegger felt that Jünger's works of the 1930s contained one of the most essential interpretations of Nietzsche's metaphysics – an interpretation so effective that it could bring us straight to the very heart of the present. According to Heidegger, the notions in *Total Mobilization* and *The Worker* give us a glimpse of the historically documentable dominion of the desire for power that by now had come to acquire a planetary dimension in the form of technology.

It is common knowledge that Ernst Jünger and Martin Heidegger[1] had the opportunity to discuss their respective theories on the general topic of nihilism "almost face-to-face", and the expansion of technology was an essential part of the debate.

[1] On the relationship between Heidegger and Ernst Jünger see, above all, the delightful essay by Franco Volpi, *Itinerarium mentis in nihilum*, which appeared as an introduction to *Über die Linie* (Adelphi, Milano 1989, pp. 11–45). This brief but concentrated work also provides all the bibliographical references relating to the questions raised by the two authors. Among the most relevant works on the topic of technology in Heidegger, see: *Kunst und Technik. Gedächtnisschrift zum 100. Geburtstag von Martin Heidegger*, edited by W. Biemel and F.W. von Herrmann, Klostermann, Frankfurt a. M. 1989; M.T. Pansera, *L' uomo e i sentieri della tecnica. Heidegger, Gehlen,*

F. Grigenti, *Existence and Machine*, SpringerBriefs in Philosophy,
DOI 10.1007/978-3-319-45366-8_5

The opportunity was provided by a volume to celebrate Heidegger's 60th birthday that was published in 1950 by Gadamer (although the editor's name does not appear in the book). Ernst and Friedrich Georg Jünger both contributed an essay to this miscellany. The contribution from Ernst was entitled *Über die Linie* and was explicitly dedicated to the topic of nihilism. Heidegger commented on it in a letter entitled *Über "Die Linie"* that appeared in the volume *Freundschaftliche Begegnungen. Festschrift für Ernst Jünger zum 60. Geburtstag* (Klostermann, Frankfurt a. M., 1955). It is worth mentioning here that the stance taken by Heidegger seems to come closer to Friedrich Georg's disenchanted pessimism than to Ernst's titanism, but this will be discussed in more depth in a dedicated study.

In his essay *Die Frage nach der Technik (The Question Concerning Technology)*, Heidegger defined technology as a modality of what he called "revealing" (*Entbergen*). According to Heidegger, all production, even if it involves the use of technical instruments, must be seen as a way in which truth, by revealing itself, becomes manifest. The problem lies in establishing whether this determination helps us to understand not only man's handiwork, but also the new phenomenon of modern technology. Heidegger himself asks the question:

> In opposition to this definition of the essential domain of technology, one can object that it indeed holds for Greek thought and that at best it might apply to the techniques of the handcraftsman, but that it simply does not fit modern machine-powered technology. And it is precisely the latter and it alone that is the disturbing thing, that moves us to ask the question concerning technology per se. It is said that modern technology is something incomparably different from all earlier technologies because it is based on modern physics as an exact science. Meanwhile we have come to understand more clearly that the reverse was true as well: modern physics, as experimental, is dependent upon technical apparatus and upon progress in the building of apparatus. The establishing of this mutual relationship between technology and physics is correct but it remains a merely historiographical establishing of facts and says nothing about that in which this mutual relationship is grounded. The decisive question still remains: of what essence is modern technology that it happens to think of putting exact science to use?[2]

It is worth taking certain parts of the above passages point by point: (a) determining technology in the light of what the Greeks meant by *poiesis* may not suffice for us to grasp what technology has become in the modern world; (b) such a determination seems to be at ease with the model of handcrafts, in contact with which it has historically been used, but it may not serve our purpose when it comes to understanding technology founded on machines for providing motive power; (c) there is a reciprocity in the link between machine and *episteme*: advances in the latter are enabled by a massive use of devices and instruments, and every theoretical gain determines an improvement in the former.

Of course, if (a) were true, it would be difficult for Heidegger to keep modern technology on the horizon of truth, in the sense of "revealing", of which *poiesis* is a

---

Marcuse, Armando, Roma 1998; N. Russo (ed.) *L'uomo e le macchine. Per un'antropologia della tecnica*, Guida Editore, Napoli 2007.

[2] Martin Heidegger, *The Question Concerning Technology and Other Essays*, trans. W. Lowitt, New York: Harper and Low 1977, pp. 13–14.

part. He also perceptively acknowledges that (b) harnessing motive power in mechanical devices represents something completely new that cannot be predicted by starting from classical distinctions that date back to technological worlds in which these machines did not exist. He claims too that (c) the relationship between science and technology should be reversed: it is not science that dominates technology, but the latter that increasingly sustains and governs the destiny of science. On the first of these questions, Heidegger says:

> What is modern technology? It too is a revealing. Only when we allow our attention to rest on the fundamental characteristic does that which is new in modern technology show itself to us. And yet the revealing that holds sway throughout modern technology does not unfold into a bringing-forth in the sense of *poiesis*. The revealing that rules in modern technology is a challenging (*Herausfordern*), which puts to nature the unreasonable demand that it supply energy that can be extracted and stored as such. But does this not hold true for the old windmill as well? No. Its sails do indeed turn in the wind; they are left entirely to the wind's blowing. But the windmill does not unlock energy from the air currents in order to store it.[3]

The verb *herausfordern* essentially means "to challenge" or "to launch a challenge". It is commonly used in expressions such as "to challenge someone to a duel" or "to challenge fate". Deriving from this, *herausfordern* also means "to provoke" or "to stimulate" something, or someone, in a particularly energetic manner. It could be translated into English as a "challenge", instead of a "challenging", but this converts into an "object" what was clearly conceived as a type of action, a way of behaving. If Heidegger had wanted to use a noun form, he would have written *Herausforderung*, but then he would certainly have lost the practical-transitive value of the verb that he evidently wished to retain.

His second consideration is even more noteworthy. The participle *herausgefördert* appearing in the German original of the above quotation is not *strictly* the past participle of *herausfordern*. The umlaut obliges us to link it to another, very similar verb, which is *fördern*. Among the meanings of the latter ("to favor", "to encourage", "to increment"), there is also a technical term meaning "to extract", "to excavate", "to bring to light", that is used in the mining industry to indicate precisely the process of "bringing forth", of taking minerals out of the ground. What is more, the first practical example of modern technology that Heidegger uses to illustrate what he means refers specifically to the activity of coalmining. So translating *herausgefördert* as "extracted" and "stored" seems particularly appropriate, because it refers without a doubt to the technological phenomenon that the philosopher actually had in mind when he used the term.

We can rule out the possibility of Heidegger having confused two verbs in his mother tongue, even though they are virtually identical in terms of their spelling. It is more likely that he wished to use both in their respective semantic contexts, exploiting their phonetic assonance, which almost suggests a common root, in which the idea of *provoking by challenging* coincides with the idea of *extracting by bringing forth*. This must have been what Heidegger had in mind, together with another idea that we shall soon see below.

---

[3] Ibid.

Once we have made sense of all Heidegger's etymological considerations, it is important to emphasize the apparent difficulty in which he finds himself when we consider the meaning of the participle *herausgefördet*. The point is: can we really consider mining activities as being characteristic of modern technology? Methods for extracting minerals from the soil have been known and used ever since ancient times. Surely the need to obtain resources from underground for use as raw materials was already contained in the concept of *poiesis*. How does mining using modern technologies differ from the extraction technology of the ancients?

It is difficult for Heidegger to evade this objection, though his position might be further analyzed by returning to a few significant references to mining activities that crop up here and there in his writings. Repeated references to the exploitation of coal deposits, to the pumping of oil from underground, and to hydroelectric power stations on rivers make us see that, when Heidegger wants to characterize the "bringing forth" demanded by modern technology, he alludes not to the raw material for use in some crafting process, but to the *energy* contained in the elements. If we return to the example of the windmill that appears at the end of the last quotation, we can see the importance of the difference that Heidegger very neatly wishes to underscore. It is one thing to use the mechanical force transmitted by the windmill's sails to turn a potter's wheel, but quite another if we use the same device to transfer and store the wind's energy in a battery. In the former case, the working circuit ends in a product. In the latter, the stored energy can be made available for an indefinite number of other uses.

In this extraction of the pure and simple energy element, Heidegger saw unfold a way of challenging the world and its resources that revealed something more resembling the essence of technology, something that the concept of *poiesis* had not yet called upon or brought to light. The enormous heat potential deriving from the combustion of coal identifies the new leading character on the technological stage:

> But, now, in order to what is coal put, for example, the putting coal on coaldfield? It doesn't put as the pitcher on the table. Just like Earth ground with respect to coal, so the coal, who in turn, is put, rather originated, in order to heat, that is already steam supplier, whose pressure activates the mechanism that keeps on a diet a making machinery factory who produces tools by which other equipments are prepared and maintained.[4]

What "challenges" coal, by extracting its thermal energy, is the *machine*. The mineral lying underground is moved (*gestellt*) for the machine; then, thanks to the power of steam, the machine moves (*stellt*) other machines that in turn move devices, … and so on, from one machine to the next.

Having reached this point, in addition to clarifying the concrete grounds supporting Heidegger's thinking (a circumstance that makes us appreciate his attention to linguistic abstraction even more), we are also in a better position to interpret the other term the philosopher uses to indicate the essence of technology more accurately. This is the noun *Gestell* (*frame*). One way to arrive at this word is by going through the dense network of semantic shifts that Heidegger himself followed

---

[4]Id., *Der Gestell,* in Bremer und Freiburger Vorträge, Klostermann, Frankfurt am main, 1994, p. 57 (Eng. trans., by Positionality o The Ge-stell, in the Bremen and Freiburg Lectures

up from one lemma to another. Starting from *herausfordern*, it is not difficult for a German-speaker to hear the verb *fordern* echoing with the meaning of "to request" or "to demand" something from someone. But "to present a request" or "to petition" are senses of the core idea in "to challenge" (*herausfordern*) that already resounds in the verb *stellen*. Heidegger himself clarifies these relationships:

> What does it mean *stellen*? We know this word by expression as "to put in front", "to represent" (*vorstellen*) something, "to put here", "to produce" (*herstellen*) something. Nonetheless we have to refrain forbelieving that our thought is already up to the simple and barely considered range of this expressions. [...]Stellen means now to challenge, to require, to force to present itself. This *stellen* happens inasmuch it is *Gestellung*, "obligatory presentation", "obligation to present itself". In the command for the obligatory presentation, it addresses men. However, among what it is present as a whole,man is not the only one present-being forced to present himself. A region is *gestellt*, "taken as a target", for coal and surfacing metal-bearing mineral. Probably the stone-surfacing is already represented in the horizon of such a *stellen* and beside it is representable only depending on it. The stones-surfacing and as such already taken into account in the perspective of (*self-present*) presenting-themselves (*sichstellen*), are provoked. The earth is summoned in such a *stellen* and she is assaulted by it. She is commanded (*be-stellt*), that is forced to present itself. It is so that we understand now and hereafter the word *bestellen*, "to order". By this commanding the land changes itself in to a coalfield zone, the soil on mineral vain. This commanding it is already different with respect to the oneused in the past by the farmer to grow his field. The work of the farmer does not provoke the plot, but he commits the seeding to the power of growth, defending it while she is rooting. In the meantime, however, the soil-working too is converted to same the commanding that gives out air to nitrogen, soil to coal and metal-bearing mineral, mineral to uranium, uranium to atomic energy and this last to a destruction who can be commanded. The farming is today mechanized food industry, that in its essence is the Same (*das Selbe*) of corpse production inside the gas chambers and in extermination camps, the Same of stop and starving of entire nations, the Same of hydrogen bomb making.[5]

Without listing them all again, it is difficult to summarize the multiplicity of senses that this passage recalls. The most unifying part concerns the unappealable nature of the *order* (*Bestellung*) in the way in which technology challenges the world. Something lying relatively well hidden – the resource – becomes a target and is summoned to appear. And once this order has been obeyed, nothing remains as it was before. The resources are converted and set to other uses in a chain of actions just like the moving parts of a mechanism that can be made more and more complicated, and adapted in a great variety of ways.

We might wonder about the motives for ordering the Earth to appear this way. What is the purpose of extracting the oil from underground? The answer that we would probably all give is, "in view of the profit that can be gained from it". But, according to Heidegger, what makes technology issue this order (*Bestellen*) has nothing to do with profit; it always aims only for another orderable entity. It is not economic interest that drives processes of technological expansion – if anything the former is driven by the latter. Within these processes, the ordering action rapidly passes from one orderable entity to the next, because everything has already been

[5] Ivi, pp. 55–57.

placed in an endless chain of actions that generates nothing more than the pure and simple orderability.

As we know, Heidegger uses the term *Gestell* (*enframing*) to describe this unending sequence of actions. In German, *Gestell* specifically means "support", "pedestal", "shelf", "frame", "armature". Clearly, these terms all refer to mechanical supporting structures, i.e. to devices that are not particularly complex, but they serve fundamental purposes such as supporting or containing in the broadest possible sense, so we might also translate *Gestell* as "assembly" or "framework". Whatever word we choose, we cannot deny that, with *Gestell*, Heidegger attempts to express the essence of modern technology in the way that it reveals its "mechanicality". The assembly is not just a device among others, but the horizon on which all the resources of the technological world, including human beings, are collected and coordinated.

The object touched by the machine is thus converted into a "standing-reserve" (*Bestand*), or "stock", something to be extracted, converted and consumed. The products are also treated as warehouse stocks: they cannot be left in their *essence as something indefinitely durable*; they must be used, exhausted, consumed like any other resource. This manner of handling the object seems to be the exact opposite of *poiesis*, in which the productive element prevails. Then there is the matter of human beings. According to Heidegger, what we call "ordering" is not something that only humans have the power to do, even though we are involved in the action of ordering. This is what gives rise to our ambiguous relationship with the machine. On the one hand, we act as the machine's accomplices in governing the world, while on the other we are really only supporting actors and can be replaced at any time. The essence of technology does not configure a human way of operating, so humans can disappear from the active side of the machine and become a mere disposable resource. Science meets with the same destiny: there is no longer any room for a pure theory that might subsequently be applied to technology. Inasmuch as it is "calculus" that can be used in a system for the storage and mechanical processing of information, scientific knowledge seems to be wholly under the dominion of the machine, and has been so ever since its origins.

It is easy to see why Heidegger repeatedly insists that the essence of technology is a technical entity or, better still, a machine; and his thinking has made a formidable contribution to the way in which we understand the specific nature of devices.

Chapter 6
Arnold Gehlen – Inadequacy and Technology

Abstract In this chapter is discussed the "basic anthropology" of Arnold Gehlen. In this author the idea of the machine is connected to: (a) the concept of *Mangel, which* expresses not the mere *absence* of something, but a more complex condition of *poverty* and *inadequacy*. The organic deficiency relates to the being of the thing, rather than to any non-presence of certain properties. It is intended to reflect the core idea of *non-specialization* (*Unspezialisierung*); (b) the idea of Man as being not designed for any specific natural environment. He must constantly adapt to every living world he happens to encounter in nature. He relies primarily on a self-referential circuit of actions which, in the particular case of technological manipulation, takes the form of *compensating for or substituting the missing organ*. It is through *integration, intensification* and *facilitation* that tools and machines compensate for man's natural organic shortcomings. In Gehlen's anthropology machines are simply the exact reflection of our weaknesses, a sort of *nature artificielle.* The *promotion/destruction of life* by the great *technological man* is apparent from the increasingly.

Keywords Gehlen • Man • Inadequacy • Organismus • Compensation • Protesis • Nature artificielle

In his major work, *Der Mensch. Seine Natur und seine Stellung in der Welt* (*Man, His Nature and His Place in the World*) Arnold Gehlen does not seem to refer explicitly to technology as yet. The great anthropologist focuses mainly on the concept of *action*[1] (*Handlung*), but Gehlen's use of this term gives us the impression

[1] See K. S. REHBERG, *Arnold Gehlen's Elementary Anthropology, Introduction to* A. GEHLEN, *Man, His Nature and His Place in the World ....* There are numerous contributions on action in Gehlen. Among the most recent in the Italian language the: U. FADINI, *Il corpo imprevisto. Filosofia, antropologia e tecnica in Arnold Gehlen*, Milano 1988; M. T. PANSERA, *L'uomo. Progetto della natura. L'antropologia filosofica di Arnold Gehlen*, Roma 1990; B. ACCARINO (ed.), *Ratio imaginis. Uomo e mondo nell'antropologia filosofica*, Firenze 1991; R. TRONCON, *Studi di antropologia filosofica. I. La filosofia dell'inquietudine*, Milano 1991; there are also some important pages dedicated to Gehlen by U. Galimberti in his recent volume *Psyche e Techne*, Milano 1999. There is a greater abundance of literature in other languages, though the critics' interest often focuses more on topics such as Gehlen's social and political views and the question of technology. As concerns action, see in particular: C. HAGEMANN-WHITE, *Legitimation als*

F. Grigenti, *Existence and Machine*, SpringerBriefs in Philosophy,
DOI 10.1007/978-3-319-45366-8_6

that, in speaking of action in general, what he really means is a number of practices, and the typically technological action is certainly among them.

Drawing from Max Scheler, Gehlen believes that the purpose of philosophical anthropology is to define the *particular place of man in the world* once and for all. This is a mission, as Gehlen points out more than once, that has been repeatedly undertaken but never truly brought to completion. The author acknowledges two difficulties that every effort made so far has failed to overcome. On the one hand, there is the problem of piecing back together the *outside* and the *inside of man*, morphology and psychology, somatic and psychic. On the other, there is a persistent tendency to consider single traits or qualities as defining features of what makes humans different. Gehlen approaches both these issues with his *elementary anthropology* that rejects all abstract or partializing definitions and places the focus instead on the real *conditions of human existence*:

> Put your mind to this unique and incomparable being which lacks all conditions of animal life and ask: in spite of this, whattasks does this being faceif it simply wants to save its life, save its own existence, last in its mere being here. Man is an animal that has not been defined yet and is somehow still not completely understood in a final way. He is, thus, as I've said, a being who finds himself through his tasks [...] that his life becomes its own task and its own business; in elementary words: it is already quite a challenge for him to be alive the following year.[2]

Seen from Gehlen's perspective, man shows very clearly that he is a particular project of nature. Given his somatic constitution, even simply coming into being and staying alive seem to constitute a mammoth task, and he has to mobilize all his resources to succeed. Hence the different relationship that man is obliged to establish with his available organic resources. While an animal simply exists by virtue of its perfect structural determinateness, man must organize and exploit his capabilities as if they were external instruments to be guided and directed. For man, it is not simply a question of living, but of *actively conducting himself* in life.

According to Gehlen, man's task (*Aufgabe*) in life is to equip himself (*verfügen*) and conduct himself (*verhalten*), both verbs that clearly express a reflexive activity (in the English language too). So the German terms for simply *placing* (*fügen*) and merely *holding* (*halten*) would not suffice; we need to speak of man *putting himself in order* (*ver-fügen*) and *keeping himself in order* (*ver-halten*). It is worth emphasizing that the prefix "*ver*" is, by definition, inseparable from the rest of the word. It is not just mechanically tacked onto the syntagma, as if it were a mere accessory, an inessential condition. It forms an indissoluble whole with the latter. Man's task cannot be identified as the sum of the animal's tasks; it constitutes a *totality of meaning irreducible to the animality*. Taking a perspective that implies no escape from the material to the spiritual, but rather a more authentic biological approach to human beings, this means observing man's higher cognitive functions – such as imagina-

Anthropologie. Eine Kritk der Philosophie Arnold Gehlens, Stuttgart 1973; P. JANSEN, *Arnold Gehlen. Die Anthropologische Kategorienlehre*, Bonn 1975; W. OSTBERG, *Sprache und Handlung. Zur frühen Philosophie Arnold Gehlen*, Diss., Tübingen 1977.

[2] Ivi, p. 17 (our translation).

tion, language, and intellectual thinking – in their *concrete occurrence,* i.e. as vital needs inseparably integrated in the living human body.

According to Gehlen, the global, innovative project that man represents in the world of nature consists primarily in the development of an organic structure that is clearly *lacking* in highly-specialized organs, i.e. it is not specifically suited to a given *environment (Umwelt).*[3]

In Gehlen, the concept of *Mangel* – generally translated as *deficiency* or *shortage* – expresses not the mere *absence* of something, but a more complex condition of *poverty* and *inadequacy*. The organic deficiency relates to the thing's being rather than to any non-presence of certain properties. It is intended in terms of the core idea of *non-specialization* (*Unspezialisierung*), which can in turn be translated by the concept of *primitive* (*primitiv*):

> In other words, morphologically–unlike all superior mammals – man is in principle characterized by a series of lacks, which can from time to time be defined with a definite biological meaning of non-adaptations, non-specializations, primitivism, that are lack of growth and so with an undoubtedly negative meaning. He lacks a fur lining, consequently he is exposed to the inclemency of the weather. He is lacking in natural defence organs, but also in a somatic structure suited to avoiding danger; as for sharpness of senses he is exceeded by most animals andhe is lacking in genuine instincts beyond the limits of peril to hisown life. As anew born and during his entire childhood he requires constant care. In other words: in original environments, finding himself earthbound, in the midst of most dangerous hunters and more than capable of seeking safety in flight man would have been eliminated from the face of the earth since time immemorial.[4]

Clearly, therefore, from Gehlen's point of view, man should be considered as a being who, ever since very remote times, has carefully avoided taking any path leading to specialization, and whose organs consequently have not lost that *fullness of opportunity*, or *undefined plasticity*, characteristic of every instrument unsuited to any particular purpose. Gehlen analyzes a whole array of instances of *organic primitivism* (*Organprimitivismen*) in man that would confirm his nature as an *embryonal being*. The shape of the human skull, especially in the relationship between the cerebral and frontal portions, the evidently undifferentiated dentition, and the anatomical structure of the hand and foot are all indicators that point unequivocally to man's destiny as a very archaic being who has never embarked on the path of adaptive improvement taken by the anthropoid apes (from which humans do not descend); instead he has retained a conformation that is *extremely* underdeveloped.

On this specific point Gehlen grafts an interesting reflection on Lodewijk Bolk's theory of *retardation.*[5] The great Dutch anatomist identified a lengthy series of primitivisms that he interpreted as *fetal states or conditions that had become permanent.* According to Bolk, such anatomical features as orthognathism, hairlessness,

[3] For the distinction between *environment* (*Umwelt*) and *world* (*Welt*), Gehlen refers to the very important work by J. von Uexküll, entitled: *Umwelt und Innenwelt der der Tiere*, Berlin 1921.

[4] Ivi, p. 34 (our translation).

[5] In particular, Gehlen quotes the following works by Bolk: le *Vergleichende Untersuchungen an einem Fetus eines Gorilla und eines Schimpansen*, "Zeitschrift für Anatomie und Entwicklungs-Geschichte", 81, 1926; and *Das Problem der Menschwerdung*, Jena 1926.

depigmentation of the skin, hair and eyes, the shape of the ear lobes, the epicanthus, the central position of the occipital foramen, the conspicuous weight of the brain, and many others, are all transient qualities or morphological conditions in other primates, but for some reason in humans they have become *so stable as to become permanent.* Successive stages of fetal development that apes grow out of thanks to a series of subsequent particular specialization steps are fixed and stable in man as part of a permanent picture of genuine developmental inhibition. The essential trait of human beings would thus be represented by the *definitively fetal configuration of his bodily form.*

Bolk identifies a specific biological fact as being responsible for this evident persistence of embryonal traits in the human body right into adulthood, and that is an organic process that he calls *developmental retardation.* This takes shape in characteristics that are exclusive to our species, i.e. an abnormally slow growth rate and the persistence of a lengthy, purely somatic life long after our reproductive function has been exhausted. According to Bolk, this general developmental retardation is rooted, in all probability, in a particular way of functioning of our *endocrine system.* Proof of this assumption would derive from the fact that impairments or diseases affecting this system soon give rise to malformations of a developmental nature, such as hirsutism, premature closure of the cranial sutures, and enlargement of the jawbone. In other words, removing the retarding and inhibiting elements would prompt the recovery of a normal rhythm of somatic development and the consequent appearance of ape-like features of specialization.

It is not that man can *return to being* an ape, he can *evolve into* an ape, losing all those primitivisms that are simply *fetal states that have become permanent.* From this point of view, man appears as a being in obvious contrast with the trend of biological evolution, which adapts highly-specialized organic forms to equally well-defined and specific environments. Every species has a morphology perfectly suited to just one *habitat.* Man seems instead to have no particular adaptation. He has a set of non-specializations that, from the biological standpoint, are archaisms or stably immature traits (as explained earlier). We only need to look at him to see that man is wholly incapable of living naturally in a given environment. He expresses an unequalled "world-openness", but this becomes a *burden* (*Belastung*) for him and a source of fatigue that is entirely unknown to animals. He is submerged by an overabundance of impressions that he must somehow succeed in governing. He has before him not a distinct *environment* (*Umwelt*) in which the meanings are already clear and instinctively obvious, but a *world* (*Welt*) or, in other words, *a potentially unlimited field of surprises* with an unpredictable structure. This field has to be processed, it has to be experienced with circumspection, taking appropriate action and countermeasures as the case requires. Man has to:

> To get himself exempted with his own instruments and actions, that is to transform the insufficient conditions of his life into a chance of surviving. […] This principle (the principle of exemption) is the key to realizing the structural law that manages all human performance…the key point is that all deficiencies of the human constitution, in its natural state,

in its animal state as it were, are a heavy burden on its survival but they are changed by man, through work on himself and effort, into tools for his life.[6]

His lack of organic specialization makes man a being open to uncontrolled and multiple stimuli against which he can oppose no specific defenses. His lack of body hair, for instance, exposes the surface of his body to a great variety of natural stimuli, such as cold and heat, dry and wet climates, light and dark, as well as cuts and bruises, and various other lesions. No animal is so receptive to environmental stimuli: all animals have a constitution adapted *a priori* to a given *habitat*, preventively protecting it against "unwanted" pressures. Fur protects against the cold and animals that have a fur coat need not worry about finding shelter when the ambient temperature drops even to very low levels. For humans, even the most apparently insignificant changes in the weather can pose a problem. Man's extremely sensitive body means that his every living act requires a nearly intolerable effort, and he has to take appropriate action and find means of protection in order to survive.

On the one hand, he must take direct action to adapt the world around him and, starting from his own shortcomings, he carefully elaborates a hierarchy of his own capabilities and establishes a stable order of expertise. In other words, human activity is undertaken in two main directions: to produce an artificial world, and to incorporate a system of cognitive, physical and emotional habits that, starting from an original plasticity, ultimately constitutes a highly-refined system for the control and subordination of all vital actions. The two directions are inseparably connected. The one cannot take place without the other, and both stem not from external influences, but from autonomous movements that acquire a *disclosing, appropriative* and *exhaustive* value. Right from infancy onwards – which reproduces on an individual (ontogenetic) level the same undetermined condition that is characteristic of our species on a phylogenetic level – everything is seen, touched, moved, treated and processed in a series of manipulations, the end result of which is a genuine adaptation (*Bearbeitung*) of the world in the sense of its usage and consumption.

By means of this process, the environment is inadvertently enriched with a very high degree of *object symbolicity*. In a continuous exchange between the eye that effortlessly sees and the hand that practices its characteristic variety of movements, man completes a work of orientation that ultimately reduces the uncontrolled flow of impressions to a limited set of objects of attention that can be readily dominated even by the gaze alone. This is how the human world of things emerged completely from the indistinct chaos of its infantile representations.

Essentially connected with man's above-described job of becoming oriented in the world, there is also a particular activity for organizing his gestures and making them more specialized. Man's organic non-specialization not only exposes him to a tempest of impressions, it is also the source of his plasticity and boundless capacity to adapt his motor responses to the surprises he might encounter in his environment.

---

[6] Ivi, p. 34 (our translation).

According to Gehlen, this incomplete motor development brings us to the second task for the survival of our species. Man has to develop his own gestural abilities, which can then be established permanently through training. This involves constantly practicing with a view to correcting errors and confirming successes in an extremely tiring effort to continuously better himself. His performance relies essentially on two features: a strong degree of *object sensitivity* in relation to the outside world, and an equally strong *self-awareness of his own movements*. The human motor system is constantly *sensitive to the touch* (*tastempfindlich*) and, with every possible movement it performs, it simultaneously receives feedback on the changes it makes. *Man not only takes action, but also feels and sees himself taking action.*

Gehlen pursues his analysis of this primary, elementary structure of human action in various directions, one of which is that essential relationship with the instrumental handling of things. And here the technological world also emerges as an opportunity for escaping, or at least easing the burden of otherwise intolerable living conditions.

Man is not designed for any specific natural environment. He must constantly adapt to every living world that he happens to encounter in nature. He relies primarily on a self-referential circuit of actions that, in the particular case of technological manipulation, takes the form of *compensating for or substituting the missing organ*. This is what Gehlen has to say on this concept:

> The oldest evidence of manual labour actually relates to weapons, which don't exist as an organ, to which should also be added the use of fire, which likewise became popular as a safety measures or thermic insulator. Present since the beginning alongsidethis principle of absent organ replacement has beenthe enhancement principle: a stone taken in one's hands to strike has a far greater effect than abare fist. So alongside integration techniques, which find a replacement for abilities ruled out by our organs, intensification techniques yield effects beyond our natural abilities: the hammer, the microscope, the phone do no more than enhance human aptitudes. Finally, there are facilitation techniques, aimed at lightening the labour of our organs, by freeing it and generally enabling them to save labour, as a wheeled vehicle makes it unnecessary to drag heavy objects by hand. Airplane travellers have the three principles in one: the plane replaces the wings which nevergrew, it certainly beats all organic flight techniques and it economize on once inevitable exertions for those who want to travel far away[7]

It is through *integration, intensification* and *facilitation* that tools and machines compensate for man's natural organic shortcomings. Structural elements that do not exist on the human body (but might be attached to it) are substituted with artificial prostheses capable of performing a function already existing in nature. Where man's performance is poor, it can be intensified beyond his normally achievable limits, but not completely modified. Then there are the facilitation technologies that can alleviate the burden of fatigue and exhaustion due to the limited resistance of man's biological body. The most advanced devices combine these three benefits into a single machine, and that is why they sometimes appear so unnatural, even monstrous. But this is only an impression. Gehlen's anthropology leads us to believe the exact

[7] A. Gehlen, *Die seele im technischen Zeitalter*, Rowohlt Taschenbuch Verlag, Hamburg 1957, pp. 12–13 (our translation).

opposite: machines are simply the exact reflection of our weaknesses, a sort of *nature artificielle* – and the farther they are removed from virgin forms and materials, the more they reinforce the shortcomings for which they must compensate:

> Despite its simply incredible brilliance, and indirect connection with nature, technology truly mirrors human beings: to be convinced it is enough to think that the oldest inventions, the essential discoveries, are not imitations of models existing in nature [...] So the world of technology is, so to speak, 'the super man'; genius and wealth of intelligence, both promoter and destroyer of life like man himself and likethe latter in multiple connections with pure nature. Technology is also, like man, *nature artificielle*.[8]

The *destruction of life* by the great *technological man* is apparent from the increasingly massive use of the inorganic to replace the organic. In an ever increasing progression, living matter is replaced by synthetic products and the power of natural organs by artificially-generated energy. While it is true that coal and oil are fuels of organic origin, the way in which they are consumed no longer bears any relationship to the goals of nature.

The same applies to all the other components of the artificial world. The processing of metals, bronze and iron, has not only led to transformations of engineering type, it has also enabled us to reach and step over new thresholds of civilization. Chemistry has subsequently given rise to the most extensive medicalization of life on synthetic grounds that humanity has ever even been able to envisage. Every technological step taken in the history of man has been achieved primarily as a form of emancipation from one or more organic substrates, and the final step will move in the direction of surpassing the body as a mere biological object.

Gehlen suggests that what enabled such a huge acceleration towards a totally artificial conception of life was not the replacement of *tools* with *machines*, but the fusion between man's knowledge of nature (seen as a dead and purely material sequence of events) and his constitutive manipulatory skills. Up until very recently, knowledge could be cultivated in a dimension entirely distinct from (and sometimes even in contrast with) any objective needs, but with the advent of scientific experimentation has come the marriage between the mechanical and the theoretical that a lengthy tradition had kept carefully separate.

The devices that scientists use (and that they often fashion themselves) are proper machines that certainly produce no economically useful results, but they enable us to fix pure natural phenomena clearly, abstractly isolating them from our still confused and imprecise experience of them. Galileo's inclined plane was the first simple machine of this type: it brings to the mind's eye a universal pattern, undisturbed by any irregularities, of a causal process that, in this form, becomes a product that is "available" for use in all kinds of applications. If there is a qualitative difference between ancient and modern technologies, it lies – according to Gehlen – in a process of reassembly between the eye and the hand such that:

---

[8] Ivi, p. 13 (our translation).

> Technology has absorbed the dynamic rhythm of progress from the new natural sciences; the natural sciences, in turn, have absorbed a more functional, constructive and non-speculative character.[9]

Clearly, Gehlen does not think it particularly important to see whether there has been any priority between theory and practice in the consolidation of modern science. What counts is the fact that, after centuries of non-innovative repetitions, a new vertical rise of human civilization has occurred thanks to the reconstitution of the basic anthropological measure: *prescient handling*.

[9] Ibidem (our translation).

Chapter 7
Günther Anders – Shame and Apparatus

Abstract In this chapter is presented the pessimistic diagnosis of Gunther Anders about the world dominated by technology. Anders observes: (a) the diffusion of a new sentiment amongst human beings, occurring in parallel with a frighteningly rapid rise in techno-science. He describes this different emotional mood (*Stimmung*) as a sense of "Promethean shame" (*prometeische Schäme*) that describes it as a hitherto unknown, but increasingly widespread sense of shame that man experiences when he sees the embarrassingly high quality of the objects that he himself has constructed; (b) the concrete realization in the present time of the by Anders is called the "mega-machine" that can be configured like a sort of huge *ideal state* in which every human and no human component serves the functional needs of the whole. According to Anders, the fusion of each particular apparatus into a single, large apparatus marks the achievement of a new ontology. From the point of view of the apparatuses, the essential meaning of things becomes the one according to which "being" means "being part of an apparatus".

Keywords Anders • Promethean shame • Apparatus • Mega-machine • Soul • Modularity • Components

From his observatory, between the United States and Europe during the uncertain times that followed immediately after the end of the Second World War, Günther Anders already saw technology as a protagonist – a disturbing leading actor that had now demonstrated, for the first time, that it was capable not only of putting man in second place, but even of cancelling him forever from the face of the Earth.

That technology had become the essential condition of human existence meant, for Anders, that it had succeeded in destroying the two pillars on which the West's spiritual and material development had been built: the presence of God and our relationship with nature. Whereas this could never have happened up until modern times, nature now disappeared from the horizon of man's main concerns. The dominion of technology over nature had radically changed its role. Nature could still be seen as a resource to consume, or as a pleasant occasional resting place, but it was no longer the term of a genuinely "formative" exchange for the members of our species.

In much the same way, technology raised radical doubts about our divine roots. In the era of maximal expansion of the machines, men not only forgot God, but –

F. Grigenti, *Existence and Machine*, SpringerBriefs in Philosophy,
DOI 10.1007/978-3-319-45366-8_7

even more culpably – they even came to disregard the fact that the very origin of technology (as we know from mythology) should be attributed to God. The power released by machines exceeds our capacity to control it because it comes from a place high above us that we cannot hope to reach. In forgetting this essential truth, man loses every reference against which to measure his real condition and comes to believe that he himself is the author and lord of technology. This misguided conviction induces him to set aside all serious critical reflection on the possible consequences of his new lifestyle, thereby exposing himself to immense dangers, even to the point of his own quite likely extinction.

In his essay of 1963 entitled *Über die prometeische Schäme*, Anders diagnoses the diffusion of a new sentiment amongst human beings, occurring in parallel with a frighteningly rapid rise in techno-science. He describes this different emotional mood (*Stimmung*) as a sense of "Promethean shame" (*prometeische Schäme*) But what exactly does he mean? Anders describes it as a hitherto unknown, but increasingly widespread *sense of shame that man experiences when he sees the embarrassingly high quality of the objects that he himself has constructed.* There is a strange reversal coming about in our relationship with the products that we manufacture:

> Therefore we have to consider the will of today's man to become a self-made man, a product, in a new context: it is true that he wants to create himself because he cannot bear the things that he is not able to make; but the real reason is that he doesn't want to be something incomplete. Not because he is outraged to have been created by others (God, gods, nature), but because he is not complete at all and, being unfinished, he is inferior to all his products.[1]

In the era of the third industrial revolution, man feels intimately humiliated not by the excellence of God, or by the sublime power of natural forces, or by the bravery of a hero, but by the *amazing structural perfection of his technological objects.* He compares himself and his organic body with the performance of which his most advanced machines are capable, and discovers – as Anders puts it – that he no longer wishes to be something not manufactured or constructed. Man is well aware that he is the product of a natural genesis that has taken place by means of processes that cannot be calculated and are difficult to fine-adjust (such as reproduction and birth). But his awareness of this fact makes him feel indignant because he realizes that he is unable to take action on his own ontic inheritance; he cannot improve it, process it, or even reconstruct it as he is able to with even the most straightforward of the appliances that he has manufactured. Very soon he begins to consider nature, from which he originates, and his own being as something inferior to even the most mundane artificial objects:

> It is indisputable that man is inferior in strength, rapidity and precision compared to his products and that his calculating machines make his intelligence seem miserable. From his point of view (derived from machines), we have to recognise that the instructor is right. This is above all because he no longer recognises his inability to carry the weight of competition

[1] G. Anders, *Die Antiquiertheit des Menschen*, Bd. I, Beck, München 2002, p. 37 (our translation).

but instead a much more modern inability: he is not thinking about man as a machine alongside other machines but about man as a machine for other machines.[2]

Set against the performance of his machines, man fails the test due to manifest incapacity. From lord of the machines, his role quickly shrinks to that of mere appendage for attaching to a device, a component serving some accessory function. Man tends increasingly to think of himself and treat himself like a machine. Hence his desire to adapt his body to give it more machine-like features, to improve its performance in every function, no longer distinguishing between the mental and the psychic spheres. Everything has to function within parameters of performativity that become ever more automated and intensified. According to Anders, this has been happening, and will continue to happen more and more with the diffusion of the potentiating practice known as *human engineering*:

Human engineering, this is the name for the auto-metamorphosis that constitutes what man tries; in other words 'engineering applied to man'. And his attempts at metamorphosis–we will specify some examples soon–consist of subjecting his physique to unusual and unnatural conditions, to situations where the body is at its limits. Not to know what the human body is like but to check the 'extreme levels'that are endurable. This is what the human engineer wants to know: not the modality of development but the abnormal stimulations that are bearable; not the insuperable limits but those bounds that are not fixed and which may, therefore, be overstepped.[3]

All attempts to convert man's body into something more closely resembling a machine rotate around the idea of removing certain functional limitations that it reveals inasmuch as it is the product of a biological development. Fatigue, a limited sensory acuity, great sensitivity to pain, emotional issues, and difficulties in maintaining high standards of cognitive performance – all such aspects are tested to the limit in the hope of being able to eliminate a human weakness or make us more efficient. In the process, man is made not more human, but increasingly similar to the machines that already perform better than him in certain specific activities.

Anders sees this not as a case of pure Promethean *hybris*, but rather as a development demanded by the levels of performativity and efficiency that the technological system demands. The body has to be retrofitted in order for it to keep pace with the performance of the best technologies in every sector. As Anders reminds us, we tend to consider machines as something stiff and dead. This view is correct as long as we are only looking at a single machine, but it does not hold for the world of machines as a whole. The latter should be seen as a system capable of changing at a mind-boggling rate, certainly far faster than the rhythms of biological evolution. As for man:

The bodies of today and yesterday are the same; today it is still our parents' body, our ancestors' body. There is no difference between a rocket builder's body and troglodyte's; in fact there is a morphological uniformity. From a moral point of view: not free, refractory and

[2] Ivi, p. 43.

[3] Ivi, pp. 48–49.

dull; from the point of view of the machines: conservative, unable to develop, antiquated, unmodifiable, an obstacle in the face of the ascent of machines.[…].[4]

Man feels that his body is chained to a morphology that is hard to change; it cannot be reshaped, and it cannot cope with the rhythm of development imposed by technology. Not all technically feasible projects are also humanly feasible, of course. Space exploration, visiting places light years away, is already a possibility for machines, but we do not know if it will ever be so for humans. And even if it is, the bodies of the first men to attempt such an enterprise will have undergone such changes that it will be hard for us to still call them "human". Man has thus begun to feel as if he is sabotaging his own efforts. Like a spanner in the works, he is vanifying the extraordinary potential of his machines, the mechanical efficiency of which can be constantly increased, affording them a freedom that no living organism will ever possess.

This explains why man feels ashamed, obsolete and inadequate when he compares himself with his machines. His body is liable to deterioration and death. His being alive seems to count for little in relation to the "immortality" that he himself can assure his products. Clearly, it is not that artificial objects can last indefinitely without suffering from wear and tear, or gradual obsolescence. The immortality of artefacts that Anders is talking about is what derives from their so-called "industrial reincarnation" or, in other words, their *existence in series*. Every single piece – be it a screw, a PC, a lamp – has its own functional durability, that in some cases can be predicted with a fair degree of accuracy. But like every mass-produced product, the light bulb can be replaced by another identical to the first that simply continues its life, without anything changing. The *light bulb per se*, the light bulb as a *type,* survives forever in every example of it that is manufactured. Anders makes the point that this technological seriality achieves a curious form of Platonism such that:

To consider an instrument in our possession mortal or immortal is purely a money matter as long as there are available spare parts, mass-produced on standard models, when we can replace a single broken piece (for example this bulb that has gone) with another one. As long a man with money exists, the possibility of reincarnation for all pieces also exists. And this possibility dies only when the concept of the piece dies, that is when its model is overtaken by another model which will take its place.[5]

Given this remarkable longevity his objects, man begins to compare his own impermanence with a world made of eternal things. He comes to realize, what's more, that the real reason for his perishability (*Sein zum Tode*) lies in his uniqueness, i.e. in the fact that it is impossible for any given individual to take the place of another. This brings us to the idea of being able to *live forever* (by being replaced by a copy) as a desirable form of existence, even at the price of losing our unique personality, something that now feels more like a burden, detrimental and limiting.

The most obvious symptom of the "immortality" sickness that has infected man since he has begun to compare himself with machines is apparent, according to

[4] Ivi, pp. 44–45 (our translation).

[5] Ivi, p. 61 (our translation).

Anders, in what he calls "iconomania". Everybody, even the most reserved of individuals without the least interest in fashion, will have at least one image of himself in his possession. Such forms of iconic reproduction give the ego the impression of somehow being able to exorcise the inherent human defect of "being only once", nurturing the illusion that the proliferation of photographs depicting him lend him a sort of multiple existence, in *more than one copy*. The power of the image.

Although man wishes more or less wittingly to resemble his machines, although he deludes himself that he can transcend the limits imposed on him by his "nature", he feels condemned to being a faulty product. This gives rise to that sentiment of shame, that Anders identifies as a particular "mood" that grips humans in the era dominated by technology.

But what is shame?

> Shame is a reflexive act (to be ashamed of oneself), so there is a relationship with ourselves; b) but it is a relationship with ourselves which fails; c) and not only occasionally (like other acts, for example memory) but a relationship which fails on principle; d) on principle because the one who is ashamed recognizes himself at the same time as identical and not identical (that is me but not me); e) therefore the reflexive act is never resolved; because those who are ashamed fail to resolve the contradiction that they find themselves in the shame does not dissolve (in this and the following two characters, f) and g), it looks like wonder); f) so that it loses its real character as an act and degenerates into a condition; g) but not into a balanced state of a stationary 'mood' but into a mood oscillating between irritation and disorientation; in a condition that seems to begin all over again, even when you believe to have been in it for a long time. So shame is a disturbance in self-identification, 'a state of perturbation'.[6]

In his relationship with machines, man finds himself in a condition of total disarticulation of identity. Switching to the other fails to achieve any recomposition of the self, leaving the individual suspended in an endless state of unachieved reflexivity. People who are ashamed of some physical defect swing constantly back and forth between having and being their own *handicap*. To the hunchback, his hump seems not something that he is, but something that he has, and that he might not have had. But our bodily failing defines our essence far more than a *proprium* from which we might be separated. We are our body and, as a consequence, we are necessarily also all that we do not like about it, and that we might delude ourselves that we can be rid of. This means that, in our relationship with our deficiencies, we swing incessantly between a state of identification and one of non-identification; we *are* and *we are not* at the same time.

The sense of shame is none other than the expression of this failure to return to the self, the effect of which is the constant displacement of man from the place he should occupy in the world.

Clearly, all this happens because there is a summons of a higher order that makes him feel obliged to answer for his own imperfection. This summons operates like a sort of tribunal that decides how humans *must* be. Once it was up to God and the laws of nature to judge human failings. Today, the summons comes from the world of technology, at the heart of which is the machine. Man grants the machine the

---

[6] Ivi, p. 72 (our translation).

faculty to decide on the anthropological criteria according to which a well-made member of our species is distinguishable from one who is not. Of course, everyone would agree that a healthy life is better than an unhealthy one (and probably that a person who is slim is "better" than one who is obese), and that a secure existence is more desirable than one spent in a state of uncertainty. At the same time, however, we know that we cannot feel wholly aligned with these criteria because they oblige us to try to improve ourselves on the strength a model that we can never fully match. The impotence of our ontic inheritance prevents us from responding whole-heartedly to the call for improvement in force in the era of the machine. We are ashamed of our body, which has failings for which we do not feel responsible, but we do not succeed in becoming machines. We can make them, we can improve them, and we can delude ourselves that we resemble them, but in actual fact we always fall back, again and again, within the inviolable limits of our organic being. Anders' prophecy has the flavor of a great intellectual challenge:

> In a certain sense we can imagine man – but it is only an idea – as being between the devil and the deep-blue sea, enclosed between two walls, bound by two forces, which together are opposed to his single existence: on the one hand weighed down by power of natural being (which belongs to the body, to the species), on the other hand weighed down by power of artificial being (bureaucratic and technical). [...] Day after day the danger increases that the I could be crushed by these two giants, having a different nature from it; millions of people hope in this disaster, this coming of a technocratic totalitarianism, and this possibility becomes more justified day after day. If this is to be the end, tomorrow or the day after tomorrow, the final triumph will be exclusively for machines; because of its greed, which pushes machines to devour everything, and above all the things most unrelated to it, machines will incorporate not only the I but also the body.[7]

Body and machine; natural and artificial; ego and id; but, above all, the insatiable hunger of the machine.

In the more mature phase of his reflections, Anders' attention focuses mainly on analyzing the category of the "world of apparatuses" (*Apparatenwelt*), giving definitive shape to the idea that the machine cannot be attributed any personal identity. In our relationship with it, we experience an entirely new mode of existence in which we are exposed to a symbiosis with an "other" that has an impersonal and anonymous role.

In the era of machines, the world itself becomes a kind of "mega-machine" that can be configured like a sort of huge *ideal state* in which every "component" serves the functional needs of the whole. According to Anders, the fusion of each particular apparatus into a single, large apparatus marks the achievement of a new ontology. From the point of view of the apparatuses, the essential meaning of things – the only one allowable – becomes the one according to which "being" means "being part of an apparatus".

Everything needs to be coordinated and brought down to a constituent part of the world's apparatus, no longer drawing any distinctions between objects, devices, people, animals, the Earth, ... and so on. The whole universe is being transformed

---

[7] Ivi, p. 86 (our translation).

into a machine. Human beings must also convert themselves into machines, or component parts of a machine. The passage to the definitive totalitarianism of the apparatuses is achieved by virtue of the traits typical of the machine, that Anders describes by listing his reasoning in ten concepts:

First thesis: machines expand themselves
Second thesis: machines' expansionist impulse is insatiable
Third thesis: the number of existing machines is decreasing
Fourth thesis: machines 'declass' themselves
Fifth thesis: machines become a single machine
Sixth thesis: the larger that mega-machine is, the more those pieces that before being reunited in it had worked as single pieces are in serious danger
Seventh thesis: in spite of the integration of pieces into the whole, each piece has to protect itself from the whole as the whole from the pieces: the pieces from the failure of the whole, the whole from the failure of the pieces.
Eighth thesis: the mega-apparatus, to which the single instruments are connected, has to give a ration of iron to all these pieces, a sort of a bridging ration, in order to support them in their role as pieces of the apparatus for as long as it, the mega-apparatus, remains in use.
Ninth thesis: one of the main tasks of all planning (that is the centralization of thousands of activities and equipment, their projection towards a single purpose) will consist, in the future, in limiting a mega-machine's size.[8]

The expansion of the machine is the primary tendency of technological evolution. Every device tends to coordinate a whole series of processes taking place outside it with its own internal functioning. The supply of power, the delivery of raw material, the collection of the product, the creation of the demand and the trends of the product's consumption all tend to proceed in step with the machine's movements. The machines themselves tend to become part of an ever larger apparatus that incorporates the original machines and any new elements associated with them.

This process can go on indefinitely, and goes to show that the machine intends not only to take control of time (as Marx had already imagined), but also and above all to occupy space, invading every adjacent territory. In this specific sense, the apparatuses have a role that is not eminently political in nature. They proliferate in the public space and spread to all the places where decisions are made. They provide resources and assure efficiency, but at the same time they phagocytose the social, offering technical and engineering solutions to problems for which the solutions should be political.

Anders has no difficulty in identifying the expansionism of the machines as the starting point of an advancing technocratic totalitarianism, already announced by the constant tendency to reduce the quantity of individual machines that succeed in remaining autonomous in the technological space. The number of single machines increases, but there is an even greater increase in the number of ways in which they can be interconnected with one another. This assembling of previously separate bodies produces more than just a link between individual entities that continue to

[8] Id, *Die Antiquiertheit des Menschen*, Bd. II, Beck, München 2002, pp. 109–119 (our translation).

retain their individuality, it creates a new macro-individual containing a new super-machine.

This explains why the number of the major devices tends to diminish. The aim of repeatedly placing things in series is ultimately to produce a single, enormous mega-machine that, in a clearly-defined hierarchy of structures and functions, coordinates the action of all the other machines that are constitutive parts of the whole. In the end, the machine reveals its monocratic tendency and its capacity to inspire thought processes characterized by a virtually exclusive call to reduce the multiplicity of the component parts and to concentrate the power of control in just one of them.

Anders emphasizes, however, that while the creation of networks of machines makes it easier to manage all the vital sectors of human existence, it also carries a new type of danger: a malfunction in even the most insignificant part of the mega-apparatus makes the whole device collapse, with potentially catastrophic consequences. Electricity grids are a good example. The degree of interconnection existing in these mega-complexes of machines is such that a blackout in a village can bring down the power supply on a country-wide scale. The damage caused by these rare, but highly destructive episodes has already induced the people responsible for managing them to invent antidotes and reduce the network's level of automation by taking action on the political relationships between single machines and the mega-apparatus. If we wish, we can interpret Anders' final considerations in the sense that we can foresee a limit to the expansion of the machine. Sooner or later, every device will stop functioning. But since we have no algorithm for deciding the point at which the individual part will "jam", even less can we hope to predict the point at which the whole system will fail. Dimensions and the number of interconnections probably define the boundaries of a machine's perfection. The globalized world, which is being built thanks mainly to machines, cannot expand beyond the limits imposed by the maximum dimensions of the technological infrastructure supporting it.

And here is Anders' tenth concept, which was missing from the previous lengthy quotation:

> Tenth thesis: characteristic of today's state of the world is not only its division into capitalist and communist emispheres, nor the fact that it is divided into technically developed and technically underdeveloped areas but also that the inhabitants of the various and very favoured regions are obliged to take a completely different position as far as technology is concerned. It would be madness, in the presence of astarving Indian whose country could be saved with the mass production of tractors, to be suspicious of technology as such. If we did, that man would have every right to fight us as his enemy.[9]

This new leap forward in Anders' attitude to technology has nothing to do with the risks that it can pose for mankind. It concerns the issue of a just distribution of the technological resources. People who complain about machines do so because they have already benefited from them. Those who have yet to benefit want machines more than anything else, but for some reason they have still not succeeded in getting

[9] Ivi, p. 119.

their hands on them. There is no reason to prevent the expansion of technology towards the parts of the planet where it can still mean the difference between life and death.

Philosophers have recently begun to discuss these issues again, and this will hopefully be a good starting point for gauging what the future holds.

Results

In the works of the authors considered in these pages we can identify three model images of the machine. These images have been drawn on at various times, also outside the realms of philosophy, and they still provide the backdrop for our knowledge of the machine, which has circulated in a great variety of languages. First of all, and especially in Marx, the machine is seen as a super-subject equipped with the capacity to compete with human beings, doing the same type of job, but far more effectively because they are more powerful and virtually immune to fatigue, at no risk of failing due to wear and tear. Man projects his own ego onto his machines, but in the form of a super-powerful being; in the practical relationships of the factory floor, the machine ultimately takes the boss's place and uses the human worker like an animate component. The idea of dominion implicated here is that of an autocracy of the artificial on economic grounds. The changes that the machine induces in human beings are all attributable to the simplicity with which the operator is connected to the machine, and to the repetitive intensification of certain serial movements that exploit only a particular part of the human body. Man tends increasingly to resemble the machine because he concentrates on developing the more mechanical characteristics of his human nature. Then Anders introduces a new sentiment in our relationship with the machines that tower above us, and that is a sense of shame, which somehow confirms in its emotional overtones that what is human is totally inferior to what is artificial.

In the other authors' writings, the machine takes on the status of a wholly impersonal and anonymous super-entity. In Ernst Jünger and Heidegger, its steel body expresses the character of an age. The machine exerts its dominion inasmuch as it is the "spirit of the time", the horizon of meaning, an acceleration and intensification that mobilizes human practices, transforming them from the inside. Everything takes on the automated traits of the machine, and even life itself tends to be superseded by mechanical perfection. But here the authors' predictions on the destiny of mankind are not entirely self-explanatory. While Jünger still considers it possible to reawaken Titan, a type of man capable of controlling technology because he has been selected by the latter, for Heidegger it is only if the epoch-making opportunity

F. Grigenti, *Existence and Machine*, SpringerBriefs in Philosophy,
DOI 10.1007/978-3-319-45366-8

unveiled by technology were to prove an illusion that there might be another chance for human existence. Man can do nothing to stop the machine, he can only accept it and allow it to govern his destiny.

Finally, in Gehlen, we find a third interpretation. Starting from anthropological considerations on man as a being with weaknesses, the machine becomes an element of that second nature that our species has been obliged to develop in order to cope with otherwise excessively difficult natural living conditions. The machine compensates for our organic weaknesses, it completes us. Our deficient structures are integrated, intensified, and facilitated, thereby ensuring the full development of what was initially a primitive and unspecialized being. It goes without saying that, in this case, the mechanical is not in opposition to the natural: for man, the machine embodies what is most natural to him, and that is the specific trait of his being incomparable and unique.

There is a unifying element detectable in these analyses, however, and that is the "power" factor, which is given at least three different meanings: as an attribute intrinsic in the machine; as a determinant of the relationship between man and machine; and as an effect of using machines on human beings.

In the first sense, the machine is powerful inasmuch as it is capable of doing more work per unit of time. The machine can complete complex sequences of movements faster than even the most expert human operator. In doing so, it interacts directly with the sources of energy that it transforms, i.e. with the fuels that it consumes in huge quantities. That is why we instinctively associate a machine's actions less with its output and more with factors such as its consumption, utilization and exploitation of resources. Second, the machine is powerful – or it might be better to say superpowerful – in its relationship with man. To (some of?) the authors discussed in this work, it is like a "being" equipped with super powers (though these powers are still comparable with those of humans), and capable of using them in opposition to man, to dominate mankind. The machine is considered hierarchically superior to human beings, and it keeps the latter in a state of inferiority, or even of servitude. In actual fact, this aspect is clearly attenuated in the anthropology of Arnold Gehlen, who instead embraces the idea of the power of the machine in the third sense, as a form of empowerment. Machines do not limit our human faculties, they are modelled on them, but they take them to a higher than normal level. The fact that they can exalt our capacity to take physical action on our environment, even to the point of enabling new and unnatural functions (such as human flight), goes to show that machines are capable of forming part of a man-machine assembly for the purpose of going beyond boundaries previously believed impossible to overcome. If there should ever be an advance that goes beyond what is human, it will happen on the horizon of the machine.

Bibliography

Main Texts

Anders, Günther. 2002. *Die Antiquiertheit des Menschen*, Bd. I, München: Beck.
Ernst, Kapp. 1877. *Grundlinien einer Philosophie der Technik*. Brauschweig: Westermann.
Gehlen, Arnold. 1940. *Der Mensch seine Natur und seiene Stellung in der Welt*. Berlin: Junker und Dünnhaupt.
Gehlen, Arnold. 1957. *Die seele im technischen Zeitalter*. Hamburg: Rowohlt Taschenbuch Verlag.
Jünger, Friedrich Georg. 1956. *Die Perfektion der Technik*. Frankfurt a. M: Klostermann.
Jünger, Ernst. 2002. Die Totale Mobilmachung. In *Sämtliche Werke*, Zweite Abteilung, Band 7. Stuttgart: Klen – Cotta.
Martin, Heidegger. 1994. Der Gestell. In *Bremer und Freiburger Vorträge*. Frankfurt am Main: Klostermann.
Marx, Carl. 1968. Das Kapital Bd I. In *Karl Marx-Friedrich Engels Werke*, Band 23, Vierter Abschnitt. Berlin: Dietz Verlag.

Secondary Literature

Accarino, Bruno (ed.). 1991. *Ratio imaginis. Uomo e mondo nell'antropologia filosofica*. Firenze: Ponte delle Grazie.
Aimone, Linda, and Carlo Olmo. 1990. *Le esposizioni universali 1851–1900*. Torino: Allemandi.
Ashby, Eric. 1979. *Scienze, tecnica e università nella rivoluzione industriale*. Torino: Stampatori.
Baculo, Adriana, et al. 1988. *Le grandi esposizioni nel mondo 1851–1900. Dall'edificio città alla città di edifici. Dal Crystal Palace alla White City*. Napoli: Liguori.
Bahr, Hans D. 1983. *Über den Umgang mit Maschinen*. Tübingen: Konkursbuch.
Baracca, Angelo, et al. 1979. *Scienza e industria 1848–1915*. Bari: Laterza.
Basalla, George. 1988. *The evolution of technology*. New York: Cambridge University Press.
Benjamin, Walter. 1986. *Das Passagen-Werk*. Frankfurt am Main: Suhrkamp 1982.
Bertho, Catherine. 1981. *Télegraphes et téléphones du Valmy au microprocesseur*. Paris: Le Livre de Poche.
Bertho, Catherine. 2000. *La démocratie et le médias au 20e siécle*. Paris: Armand Colin.
Biemel, Walter, and Friedrich Wilhelm von Herrmann (eds.). 1989. *Kunst und Technik. Gedächtnisschrift zum 100. Geburtstag von Martin Heidegger*. Frankfurt am Main: Klostermann.

F. Grigenti, *Existence and Machine*, SpringerBriefs in Philosophy,
DOI 10.1007/978-3-319-45366-8

Blumenberg, Hans. 2009. *Geistesgeschichte der Technik*. Frankfurt am Main: Suhrkamp.
Bohnsack, Almut. 1993. *Der Jacquard-Webstuhl*. München: Deutsches Museum.
Bolk, Lodewijk. 1926. *Das Problem der Menschwerdung*. Jena: Gustav Fischer.
Bolter, David J. 1984. *Turing's man. Western culture in the computer age*. Chapel Hill: University of North Carolina Press.
Bonadé Bottino, Vittorio. 2001. *Memorie di un borghese del novecento*. Milano: Bompiani.
Bonelli, Franco. 1982. *Acciaio, per l'industrializzazione. Contributi allo studio del problema siderurgico italiano*. Torino: Einaudi.
Bonnefon-Craponne, Louis. 1916. *Le Conservatoire national des Arts et Métiers*. Paris: Cnam.
Bosoni, Gianpiero, and Andrea Nulli. 1999. *L'epoca del treno, Dall'Ottocento ai giorni nostri*. Mondadori: Milano.
Bradley, Arthur. 2001. *Originary technicity. The theory of technology from Marx to Derrida*. Basingstoke: Palgrave Macmillan.
Braudel, Fernand. 1979. *Civilisation matérielle, économie et capitalisme (XVe—XVIIIe siècle)*. Paris: A. Colin.
Bredekamp, Horst. 1996. *Nostalgia dell'antico e fascino della macchina. Il futuro della storia dell'arte*. Milano: il Saggiatore.
Breton, Philippe. 1995. *L'utopia della comunicazione*. Torino: Utet.
Bruno, Jacomy. 1990. *Une histoire des techniques*. Paris: Gallimard.
Bury, John Bagnell. 1932. *The idea of progress. An inquiry into its origin and growth*. New York: Macmillan.
Cambiano, Giuseppe. 1971. *Platone e le tecniche*. Torino: Einaudi.
Capocaccia, Agostino (ed.). 1973–1980. *Storia della tecnica*. Torino: Utet.
Cardwell, Donald S.L. 1972. *Technology, science and history*. London: Heinemann.
Caserani, Gian Paolo. 1969. *I falsi Adami. Storia e mito degli automi*. Milano: Feltrinelli.
Cipolla, Carlo M. 1962. *The economics history of world population*. Harmondsworth: Penguin.
Clarck, Andy. 2007. *Beeing there*. Cambridge, MA: MIT Press.
Colombo, Umberto. 2000. *Energia. Storia e scenari*. Roma: Donzelli.
Cordeschi, Roberto. 1998. *La scoperta dell'artificiale. Psicologia, filosofia, e macchine intorno alla cibernetica*. Milano: Dunod.
D'Atri, Annabella. 2008. *Vita e artificio. La filosofia di fronte a natura e tecnica*. Milano: Rizzoli.
De Galiana, Thomas, and Michel Rival (eds.). 1996. *Dictionnaiere des inventeurs et inventions*. Paris: Larousse.
De Garis, Hugo. 2005. *The artilect war: Cosmists vs Terrans: A. Bitter controversy concerning whether humanity should build godlike massively intelligent machines*. Palm Springs – CA: ETC Publications.
De Palma, Armando. 1971. *Le macchine e l'industria da Smith a Marx*. Torino: Einaudi.
De Seta, Cesare (ed.). 1985. *Storia d'Italia. Annali 8. Insediamenti e territorio*. Torino: Einaudi.
Debeir, Jean Claude, et al. 1986. *Les servitudes de la piussance. Une histoire de l'énergie*. Paris: Flammarion.
Derry, Thomas K., and Trevor I. Williams. 1960. *A short history of technology*. Oxford: Clarendon Press.
Di Pasquale, Salvatore. 1996. *L'arte del costruire. Tra conoscenza e scienza*. Venezia: Marsilio.
Di Spirito, Fabio, et al. 1996. *Lo strabismo telematico. Contraddizioni e tendenze della società dell'informazione*. Torino: Utet Libreria.
Diamond, Jared. 2000. *Armi, acciaio e malattie. Breve storia del mondo negli ultimi tredicimila anni*. Torino: Einaudi.
Dijksterihuis, Edward J. 1956. *Die Mechanisierung des Weltbildes*. Berlin/Göttingen/Heidelberg: Springer.
Dugas, René. 1950. *Histoire de la Mécanique*. Neuchâtel: Griffon.
Dyson, George B. 2000. *L'evoluzione delle macchine. Da Darwin all'intelligenza globale*. Milano: Raffaello Cortina.
Elkana, Yehuda. 1974. *The discovery of consevation of energy*. London: Hutchinson Educational.

Ellul, Jacques. 1977. *Le systéme technician*. Paris: Calmann-Lévy.
Fadini, Ubaldo. 1988. *Il corpo imprevisto. Filosofia, antropologia e tecnica in Arnold Gehlen*. Milano: Franco Angeli.
Federico, Giovanni. 1994. *Il filo d'oro. L'industria mondiale della seta dalla restaurazione alla grande crisi*. Venezia: Marsilio.
Feenberg, Andrew. 1999. *Questioning Technologie*. London: Routledge.
Feenberg, Andrew. 2010. *Between reason and experience. Essays in technology and modernity*. Cambridge, MA: MIT Press.
Filippi, Federico. 1983. *Dall'elica al getto. Breve storia dei propulsori aeronautici*. Torino: Eda.
Flichy, Patrice. 1994. *Storia della comunicazione moderna*. Bologna: Baskerville.
Forbes, Robert J. 1950. *Man the maker*. New York: Schuman.
Friedmann, Georges. 1946. *Problèmes humains du machinisme industriel*. Paris: Gallimard.
Galimberti, Umberto. 1999. *Psyche e Techne*. Milano: Feltrinelli.
Garbrecht, Günter (ed.). 1987. *Hydraulics and Hydraulic Research. A Historical Review*. Rotterdam-Boston: A.A. Balkema.
Gehlen, Arnold. 1990. *L'uomo. La sua natura e il suo posto nel mondo, introduzione di Rehberg, K. S.* Milano: Feltrinelli.
Giadina, Andrea (ed.). 1989. *Le cose della guerra*. Milano: Fondazione Lorenzo Valla-Mondadori.
Gibson, James. 1979. *The ecological approach to visual perception*. Boston: Houghton Mifflin.
Giddens, Anthony. 1984. *The constitution society*. Cambridge: Polity Press.
Gille, Bertrand (ed.). 1978. *Histoire des techniques: technique et civilisation, technique et science*. Paris: Gallimard.
Giovannini, Giovanni. 2003. *Dalle selce al silicio*. Milano: Scheiwiller.
Goldfinger, Charles. 1996. *L'utile e il futile. Per un'economia dell'immateriale*. Torino: Utet Libreria.
Hagemann-White, Carol. 1973. *Legitimation als Anthropologie. Eine Kritik der Philosophie Arnold Gehlens*. Stuttgart: Kohlhammer.
Hermann, Armin, and Charlotte Schonbeck (eds.). 1993. *Technik und Kultur*, X Bände. Düsseldorf: UDI Verlag.
Hobsbawn, Eric John. 1984. *World of labour*. London: Weidenfeld and Nicolson.
Ingold, Tim. 1999. Tools for the Hand, Language for the Face: An Appreciation of Leroi-Gourhan's "Gesture and Speech". *Studies in History and Philosophy of Biological and Biomedical Sciences* 30 (4): 411–453.
Ingold, Tim. 2000a. Evolving skills. In *Alas poor Darwin. Arguments againts evolutionary psychology*, ed. H. Rose and S Rose (a cura di), 225–246. London: Jonathan Cape, pp. Id.
Ingold, Tim. 2000b. *The perception of the environment: Essays on livelihood, dwelling and skills*. London: Routledge.
Ingold, Tim. 2008. *When ANT meets SPIDER: Social theory for arthropods*. In *Material agency. towards a non-anthropocentric approach*, ed. C. Knappett e L. Malafouris, 209–215. New York: Springer.
Inkster, Ian. 1991. *Science and technology in history. An approach to industrial development*. London: Macmillan.
Jansen, Peter. 1975. *Arnold Gehlen. Die Anthropologische Kategorienlehre*. Bonn: Bouvier.
Joas, Hans. 2005. *The creativity of action*. Oxford: Polity Press.
Klemm, Friedrich. 1954. *Technik, Eine Geschichte ihrer Probleme*. Freiburg-München: K. Alber.
Lagrée, Michel. 2000. *La Bénédiction de Prométhée. Religion et technologie*. Paris: Fayard.
Landes, David S. 1969. *The unbound Prometeus, technological chance and industrial development in western Europe from 1750 to the present*. Cambridge: Cambridge University Press.
Landes, David S. 1983. *Revolution in time*. Cambridge: Harvard University Press.
Latouche, Serge. 1995. *La megamacchina*. Torino: Bollati Boringhieri.
Latour, Bruno. 1987. *Science in action: How to follow scientists and engineers through society*. Milton Keynes: Open University Press.

Latour, Bruno. 2005. *Reassembling the social: An introduction to actor-network-theory*. Oxford: Oxford University Press.
Le Goff, Jacques. 1986. *Storia e memoria*. Torino: Einaudi.
Leroi-Gourhan, Andrè. 1977. *Le geste et la parole*. Paris: Michel.
Mach, Ernst. 1883. *Die Mechanik in ihrer Entwicklung historisch-kritisch dargestellt*. Leipzig: Brockhaus.
Maldonado, Tomàs (ed.). 1979. *Tecnica e cultura. Il dibattito tedesco tra Bismarck e Weimar*. Milano: Feltrinelli.
Mängel, Manfred. 2005. *Das Wissen des Kriegers. Der Magische Operateur*. Berlin: Xenomonos.
Mantoux, Paul. 1905. *La révolution industrielle au XVIIIe siécle*. Paris: Société nouvelle de librairie et d'édition.
Marchis, Vittorio (ed.). 1995. *Storia delle scienze. 5. Conoscenze scientifiche e trasferimento tecnologico*. Torino: Einaudi.
Marchis, Vittorio. 2010. *Storia delle Macchine. Tre millenni di cultura tecnologica*. Roma-Bari: Laterza.
Marchis, Vittorio, and Filippo Nieddu. 2004. *Materiali per una storia delle tecniche*. Torino: Celid.
Mathias, Peter, and John A. Davis. 1989. *The first industrial revolution*. Oxford: Basil Blackwell Ltd.
McNeil, Ian (ed.). 1989. *An encyclopedia of the history of technology*. London/New York: Routledge.
Michael, Eldred. 2000. *Kapital und Technik*. Dettelbach: J.H. Röll.
Micheli, Gianni. 1995. *Le origini del concetto di macchina*. Firenze: Olschki.
Mokyr, Joel. 1997. *Leggere la rivoluzione industriale*. Bologna: il Mulino.
Musson, Alert Edward, and Eric Robinson. 1969. *Science and technology in the industrial revolution*. Manchaster: University of Manchester Press.
Musto, Francesco. 1990. *Storia della tecnologia radar: l'avventura di Argo*. Roma: Bariletti.
Nacci, Michela (ed.). 1998. *Oggetti d'uso quotidiano. Rivoluzioni tecnologiche nella vita d'oggi*. Venezia: Marsilio.
Negrotti, Massimo. 1990. *Capire l'artificiale*. Torino: Bollati Boringhieri.
Nerhot, Patrick. 2008. *Ernst Jünger-Martin Heidegger: il senso del limite (o la questione della tecnica)*. Padova: Cedam.
Noble, David F. 1993. *La questione tecnologica*. Torino: Bollati Boringhieri.
Ord-Hume, Arthur W.J.G. 1977. *Perpetual motion. The history of an obsession*. London: George Allen & Unwin Ltd.
Ortega, Josè y. Gasset. 1982. *Meditación de la técnica y otros ensayos sobre ciencia y filosofía*. Madrid: Alianza.
Ostberg, Wolfgang. 1977. *Sprache und Handlung. Zur frühen Philosophie Arnold Gehlen*. Tübingen: Dissertation.
Pansera, Maria Teresa. 1990. *L'uomo. Progetto della natura. L'antropologia filosofica di Arnold Gehlen*. Roma: Studium.
Pansera, Maria Teresa. 1998. *L' uomo e i sentieri della tecnica. Heidegger, Gehlen, Marcuse*. Roma: Armando.
Pollard, Sidney. 1981. *Peaceful conquest. The industrialization of Europe 1760–1970*. Oxford: University Press.
Pomian, Krysztof. 1984. *L'ordre du temps*. Paris: Gallimard.
Portoghesi, Paolo. 1981. *Infanzia delle macchine*. Roma-Bari: Laterza.
Reynolds, Terry S. 1983. *Stronger than a Hundred Men. A History of the Vertical Water Wheel*. Baltimore: The John Hopkins University Press.
Rosenberg, Nathan. 1982. *Inside the black box: Technology and economics*. Cambridge: Cambridge University Press.
Rossi, Paolo. 1971. *I filosofi e le macchine (1400–1700)*. Milano: Feltrinelli.
Russo, Nicola (ed.). 2007. *L'uomo e le macchine. Per un'antropologia della tecnica*. Napoli: Guida Editore.

Scheiwiller, Vanni (ed.). 1989. *Civiltà delle Macchine. Antologia di una rivista 1953–1957*. Milano: Scheiwiller.
Schiera, Pierangelo. 1987. *Il laboratorio borghese. Scienza e politica nella Germania dell'Ottocento*. Bologna: Il Mulino.
Severino, Emanuele. 1979. *Téchne. Le radici della violenza*. Milano: Rusconi.
Silvestri, Marco. 1988. *Il futuro dell'energia*. Torino: Bollati Boringhieri.
Singer, Charles (ed.). 1957. *A history of technology*. London: Oxford University Press.
Somenzi, Vittorio, and Roberto Cordeschi. 1986. *La filosofia degli automi*. Torino: Bollati Boringhieri.
Spengler, Oswald. 1931. *Der Mensch und die Technik*. München: Beck.
Strack, Gilbert. 2000. *Titan Technik. Ernst und Friedrich Georg Jünger über da technische Zeitalter*. Würzburg: Königshausen und Neumann.
Timmermans, Bernard. 2003. L'influence hégélienne sur la philosophie de la technique de la technique d'Ernst Kapp. In *Les philosophes et la technique*, ed. Chabot-Hottois, 95–108. Paris Vrin.
Troncon, Renato. 1991. *Studi di antropologia filosofica. I. La filosofia dell'inquietudine*. Milano: Guerini e Associati.
Turner, Gerard L'.E. (ed.). 1991. *Storia delle scienze. Gli strumenti*. Torino: Einaudi.
Uexküll, Jakob Johann. 1909. *Umwelt und Innenwelt der Tiere*. Berlin: Springer.
Viulac, Jean. 2009. *L' époque de la technique; Marx, Heidegger et l'accomplissement de la métaphysique*. Paris: Presses Univ. De France.
Volpi, Franco. 1989. *Itinerarium mentis in nihilum*, which appeared as an introduction to *Über die Linie*, 11–45. Milano: Adelphi.
Womack, James P., et al. 1993. *La macchina che ha cambiato il mondo*. Milano: Rizzoli.

Index

F. Grigenti, *Existence and Machine*, SpringerBriefs in Philosophy,
DOI 10.1007/978-3-319-45366-8

Made in the USA
Coppell, TX
16 January 2020

14598467R00050